HOW I RESCUED MY BRAIN

David Roland has a PhD in clinical psychology, and has trained in neuropsychological assessment and studied interpersonal neurobiology online with professor Daniel Siegel (author of *Mindsight*). David is an honorary associate with the School of Medicine at the University of Sydney, a member of the Australian Psychological Society, and a founder of the Australian branch of the Compassionate Mind Foundation. His first book was *The Confident Performer* (1998).

HOW I RESCUED MY BRAIN

a psychologist's remarkable recovery from stroke and trauma

DAVID ROLAND

SCRIBE
Melbourne • London

Scribe Publications
18–20 Edward St, Brunswick, Victoria 3065, Australia
2 John St, Clerkenwell, London, WC1N 2ES, United Kingdom

Published by Scribe 2015

Typeset in 11.5/15.5 pt Adobe Garamond Pro by the publishers
Printed and bound in the UK by CPI Group (UK) Ltd, Croydon CR0 4YY

National Library of Australia
Cataloguing-in-Publication data

Roland, David, author.

How I Rescued My Brain: a psychologist's remarkable recovery from stroke and trauma / David Roland.

9781925106008 (Australian edition)
9781922247421 (UK edition)
9781925113044 (e-book)

1. Roland, David. 2. Cerebrovascular disease–Patients–Biography.
3. Cerebrovascular disease–Patients–Rehabilitation. 4. Brain–Wounds and injuries–Patients–Biography. 5. Brain–Wounds and injuries–Patients–Rehabilitation.

362.196810092

This project has been assisted by the Australian government through the Australia Council for the Arts, its arts funding and advisory body.

Australia Council

scribepublications.com.au
scribepublications.co.uk

To Edward George Roland:
World War II pilot, telephone technician,
environmental activist,
and my father.

AUTHOR'S NOTE

IN WRITING THIS memoir, I have relied upon personal journals, medical records, and my recollections. I have consulted with most of the individuals in the story to check factual details, where these can be verified. In some cases, the individuals have helped me to re-create events and dialogue. Yet, ultimately, this is my version of the story and true to my emotional experience. Others may have different experiences of events that we shared.

I have changed many of the names and identifying details to preserve anonymity. In some cases, where individuals have agreed, I have retained real names and details.

I have not created composite characters or events. In a few instances, I have compressed time for the ease of storytelling. I have omitted events that do not relate directly to the themes in this book, and because it is impossible to include everything that happened over the six years that the memoir covers.

PROLOGUE

I'M HAVING TROUBLE working out where I am.

Somehow, this isn't perturbing, simply puzzling. I'm in a puzzle and need to put together the clues to work out what this is all about.

I'm sitting in the centre of a row of beige plastic chairs. When I turn my head, I realise that my wife, Anna, is next to me. A ring of chairs also lines the walls. Other people, scattered around the room, are flicking through magazines or looking down and shuffling their feet. I get the feeling that they don't want to be here.

A beeping sound is coming from somewhere. To my right, people are moving through an automatic door, which shudders as it opens. I look up to see a woman behind a counter and a glass window. She seems harassed, and her dark hair hasn't been brushed recently. She's on the phone and taking notes. Every so often, people go up to the opening in the window and talk to her. She frowns when they approach, as though she doesn't really want to speak with them.

Now Anna goes up and talks to her too.

We seem to be in some sort of waiting room, but I don't know why.

I look around. There are posters on the walls, with bold letters reading, 'Cover Your Cough', 'For Infection Control Reasons, Wash Your Hands', and other things. Lots of the posters have people on them looking pleased or sad.

The sunlight slanting through the windows on the far wall is soft; it must be morning light.

In one corner of the room, on a low table, there are piles of magazines. I walk over to pick up *Country Life* and then sit down again. There's a section on real estate, with pictures of quaint, homely-looking cottages; some have picket fences. Other photographs show mansions, built of sandstone or solid-looking bricks. The descriptions beneath list each property's features, telling of the life of contentment that can be enjoyed if one makes the place their new home.

There's one I like: a cute cottage with a garden for $350,000. Is that a lot of money? I used to know. When I look at the date on the cover, 2007, I realise that I don't know what year it is now. The magazine must be old: its pages are curled and creased. The names of the country towns in the ads seem familiar, but when I try to picture where they are, I can't; my sense of geography is wavy. *Goulburn*, I say to myself. Nothing. *Cooma*. Still nothing. The names swim around in my mind, sounds without any pictures attached to them.

Off to my left, a child is whining. I turn to see a man and a woman, both big, with a girl aged four or five. They look tired, as parents do when they've been up during the night with a grumpy child. Soon I am absorbed by their interactions; it's like watching a show. The father lifts the girl onto his lap, looking strained. The mother holds up a children's book, reading to her as a kindergarten teacher would. The child listens for a while, fidgets, and cries again. The mother tries to interest her in one of the toys from a box in the corner, but it doesn't work. I know what this is like; I'm a parent too. They're doing their best.

How did I get to this room? A fragment comes into my mind — a dreamlike image — of Anna driving us in the white Tarago and me vomiting out of the car window. Did this really happen, or am I imagining it?

I turn to Anna and see that she's crying quietly: her cheeks are pink; the rims of her eyes are red. She's sad about something, but I don't know what. I put my arm around her shoulder and pat her gently. 'It'll be all right,' I say. She quiets a little. After a while I take my arm back and return to *Country Life*.

As we sit there, I feel as if I'm in a sound bubble, into which the surrounding noises don't intrude. The crying girl doesn't irritate me as I think she might have at another time. Instead I feel a well of stillness inside. I keep turning the pages.

PEOPLE COME IN and out of the room, as though it becomes more and then less popular. Then a man in white appears, like a jack-in-the-box, out of a doorway. He calls out my name and holds the door ajar. It has an important-looking sign on it: CLINICAL INITIATIVES NURSE.

Anna and I get up and follow him in.

The room is small and square-shaped with clean, shiny equipment. The lights are very bright. The man has a sense of enthusiasm and energy about him; he looks interested in me. We sit down opposite each other, knee-to-knee. He brings his face, with intense, smiling eyes, close to mine. He looks clean, as though recently showered and shaved. I like his energy.

'Now, David,' he says. 'Can you tell me what day of the week it is?'

His expression is encouraging, like a teacher's. He knows the answer, but it's important to him that I say it. I want to help, so I think hard.

'It's Wednesday … or it could be Thursday.' I remember I was meant to do something special with the kids today, but I don't know what that was.

'Where are you now?' he asks.

This is harder than the day-of-the-week question. 'Is it a hospital?' It's the best thought I can come up with.

He looks satisfied. He wants to know why I'm here. I turn to Anna. She also has that knowing look, and prompts me to answer, but I have no idea. It's a mystery.

He asks more questions. I either don't know the answers or can't remember the start of the question, if it's long, by the time he's finished speaking. I'm disappointed that I can't help more. But as he talks, his words appear in my mind slowly, like tree trunks appearing out of a fog. The words often disappear before I can get hold of them, as if they are in a line, each being jostled along by the next. I'm trying to hold on to each one while he's trying to rush them. I'm feeling rattled now.

After his questions stop, he smiles and sends us out into the waiting room. The parents with the girl have gone, and most of the people are new. We must have been in the room longer than I thought. Anna must also be feeling better: I can hardly tell she's been crying.

As we wait, the stillness returns; I'm back in the sound bubble. I'm not sure, now, if the man in white was real or I imagined him. It feels as if I'm in a movie and watching it at the same time.

AS FAR AS I can tell, it's not long before we are taken through another door. It opens, like magic, into a wide, yellow corridor with a side table, a high metal chair, and shelves along the walls.

A young man who says he is a doctor asks me to sit in the chair while he stands before me. He's wearing ordinary clothes and is not enthusiastic like the man in white. Instead he looks tired, speaking slowly and softly. He probably wants to go home.

The doctor would like to know the day of the week — it seems that this is an important piece of information. Once again, I'd like to oblige, and think hard. But I get the same answer: it's Wednesday, or it could be Thursday, I say. He also wants to know where we are, and by now I know we are in Lismore Hospital because either Anna or the man in white has told me, and I've remembered. I'm confident that we were in a waiting room, because in hospitals you spend time waiting.

People come to hospitals for help. But why are we here?

And where *is* Lismore Hospital? The name is familiar, but it swirls in my mind without a picture. I have an inkling I've been here before, though. The memory's there, on the edge, just out of reach.

The doctor wants to know who Australia's prime minister is. Paul Keating's face comes to mind, but … we've had a new prime minister since Keating. Why don't I know whom? An image of a balding man with large glasses comes to mind. 'John Howard!' I say. But then, 'No, I don't think it's John Howard.' I'm unable to answer more of the doctor's questions — after he asks each one, I can't remember what he's just said. I'd like him to stop.

Now he wants to take blood — from both arms, he says, because he needs quite a bit of blood. 'Okay,' I say. Usually I'd be nervous about this, but I'm not, and offer him my left arm first. I close my eyes. There's a sensation of the needle going in, and then — nothing.

I don't know how long I'm with the doctor — perhaps one or two minutes — and when I look up he's gone. There's that puzzling feeling again: was he real, or am I in a dream?

I'M SITTING COMFORTABLY in the chair in the yellow corridor when a new man and a woman, both dressed in white, say hello.

'Oh, hello,' I say.

They tell me I am to have a CT scan. The thought excites me. I don't think I've ever had one before, but I know what they are: I've read reports from CT scans in client files, detailing the effects of a brain injury or disease.

They have a wheelchair and push me into another corridor. In and out of lifts we go. It is fun.

Now we're entering a room with a giant shiny doughnut, and a sliding platform that goes into it. My head will go into the doughnut, they say.

Before I know it, I'm being pushed away from the CT room — they say the scan is over, but I don't remember having it. How odd!

Now I'm back in an armchair in the corridor. It feels like home. I close my eyes; I'm tired. Anna is still with me, but we're not talking much.

All of a sudden there's a new man, older, with a younger man beside him. The older one must be important, because his face is serious and he's wearing a tie. Oh, they're both doctors, I realise suddenly; they have stethoscopes around their necks. The younger one must be his junior.

'Hello, I'm Doctor —,' he says, but his name slips away before I can catch it. He's standing up, looking down at me. He seems worried by something. Is it me? He also wants to know what day of the week it is and who the prime minister is, and he wants me to count backwards from one hundred by threes. I think I do all right with this one. I've always been fairly good at maths.

'What is the last thing you remember happening?' he asks.

I do remember something. 'I was playing guitar with my friend Nick. Last night.' It doesn't seem long ago.

As with the other staff, his words appear out of the fog, my answers disappearing soon afterwards. He's asking a lot more questions than the other doctor. He has a strong energy about him and I'm getting rattled again. He tells me something that seems important but I don't quite catch it. Then he's gone.

Anna has gone too. But this is okay. Something else will happen. I'll just wait.

A WOMAN IS standing in front of me, saying my name. She must be an office person: she has a penholder around her neck with a fat pen in it. She's dressed in blue pants and a spotty blouse. She gives me a clipboard with a form on it. 'This is for your health insurance,' she says.

The woman wants me to fill it in. My name and date of birth — I know these. As I go down the page, the questions get harder, and they waft in and out of the fog in my mind. I'm not sure about my answers. She wants me to sign at the bottom. My instinct says that I shouldn't sign something I don't understand, but Anna's not here to tell me what to do.

'I don't want to sign,' I say. 'I'm not sure about it.'

She nods and goes away.

Now I'm walking with a woman, also dressed in blue; she's told me that I'm staying in a ward tonight. I'm not sure if I've stayed in hospital before, but I'm so tired that I think it would be great to spend the night here. I follow her into a lift, through doors, and along corridors. We stop when she speaks to another person in blue behind a counter, this time without a glass window. The woman in blue, who I think must be a nurse, leads me into a room with three men around my age, each in pyjamas. She points to a freshly made bed. I lie down. Ah, peace and quiet.

Suddenly, without warning, there is a loud noise: wheezing and then whirring. It pierces my brain. I look to where the noise is coming from. The man in the bed beside mine is breathing into a tube attached to a machine.

I can't stay in this room with this sound. I follow my steps back, needing to think hard about which direction I came from. My sound bubble has been shattered and I feel distressed. I get to the counter. There are two nurses here now. I tell them I cannot be in the same room as the man with the machine.

'You'll get used to it,' says one of them.

'It won't be on all the time,' says the other.

Their words don't reassure me at all. They don't understand how much it hurts my brain.

'I don't want to stay here anymore,' I say.

I need to leave this place. I don't know where I'll go, but I'll catch a taxi. I walk along the corridor, away from the noise of the machine, and come to an area with lifts. I'm about to get into a lift when I notice upholstered chairs along the walls. They look soft. There's no one around and it is quiet. I'll rest here awhile before I leave.

I close my eyes and follow my breathing; my sense of calm returns. The idea of escaping slides away.

Then I hear someone come and sit down beside me. 'Hello,' a voice says. I open my eyes: it is a man in a security guard's uniform.

'Hello,' I say.

Another security guard, a bigger man, comes and stands in front of me.

'You're not going to do a runner, are you?' the first one asks.

'No,' I say, but it reminds me that I had wanted to escape. If I make a dash for the lift now, they'll catch me. The first man says something else to me, but I'm not going to answer; I'm going to be with my thoughts, my eyes closed.

I hear the second guard sit alongside the first, and they exchange a few words and laugh. Then the second guard leaves.

We've been sitting quietly for a while when my name is called. I open my eyes to see one of the nurses standing in front of me. She's smiling and says she has arranged a new room for me; an elderly patient who is going home in the morning has agreed to move into the bed next to the machine. I'm so grateful; I'd like to thank him, but when I follow the nurse back into the ward I can't remember where the first room was.

The nurse shows me the new room, and straightaway I'm reassured. The other men are elderly and seem quiet. 'Hello,' I say. Two of them respond; one is asleep.

I'm about to sit on the bed when I catch sight of the windows. Through them, a long, horizontal strip of orange is glowing, topped with purple and black. What is it? I stare and stare; I can't work it out. Then I realise: it's a sunset.

How could this be? It should be morning.

I stand and watch the orange glow become thinner and more intense as the black above it grows. The lights in the room get brighter and brighter, and begin to sting my eyes. It must be night. Incredible. Well, perhaps it will be dinnertime soon. I haven't eaten all day. Or did Anna give me a banana earlier?

The night goes by. The lights and noises are different from those at home. I sleep soundly, except when the nurse comes in to check on us and I hear low voices and rustling. When it's my turn, she apologises and shines a pin-light torch in my eyes. It stings a little. She asks me to wriggle my toes and squeeze her hand. 'I don't want to hurt you,' I say. She responds kindly: 'It won't hurt,' she says. 'Grip as tightly as you can.'

It seems that in no time at all, daylight fills up the windowpanes. I realise I'm hungry. A woman pushing a multi-level trolley brings me a tray with a small packet of cereal, stiff cold toast, and tea. It's like aeroplane food, as if I'm going on a holiday. I enjoy the breakfast, even though it's not what I'd have at home.

This morning is different from yesterday. It feels as if I've woken from a dream. I'm sure now that I'm in hospital, and that something really has happened to me. I remember more clearly the night before I came in. I'd woken with a headache, walked to the kitchen, taken a Panadol, and gone back to bed. That's the last memory I have before being here.

A nurse comes in and tells me that the specialist — the serious doctor — will be doing his rounds this morning and will discuss the test results with me. I am to stay in my room until he comes. Afterwards, I can walk around. I ask for a headache tablet.

I'm looking forward to seeing the specialist: I'm keen to know what the results say, what he thinks has happened to me. In the meantime, I enjoy getting showered, dressed, and organised. The man next to me asks what I'm in for and I tell him that I've lost my memory for some reason. I chat a little with the others and then look out the window. We are up high, and I peer down on oblong houses with broccoli trees in their backyards.

The phone beside my bed rings, interrupting my reverie. It's my psychiatrist, Doctor Banister. Anna has called him, he says. 'What's happened?' he asks.

I tell him that I can't recall most of yesterday.

'What do you think brought this on?'

I remember I'd had a huge panic attack the day before I came to hospital, after a meeting with our barrister. He'd told me that Anna and I were going to be sued.

Doctor Banister asks me what tests have been done. I mention the CT scan and the blood tests, and say I'm waiting to discuss the results with the specialist.

'You may have had a psychogenic fugue: an episode of amnesia. But we'll need to wait and see what the results reveal. I'll try and come in to see you. If it's a fugue, you could come and stay at a clinic I work for, Seaview Psychiatric Clinic, for a longer rest. I can discuss this with your doctor.'

'Okay,' I say. That does sound good.

Not long afterwards, the specialist comes in and stands by my bed, with a young female doctor this time. He looks fresh but more rushed than yesterday. He asks me how I'm feeling.

'I'm woolly in the head, as if I'm not sure I'm really here,' I say. 'I've got a mild headache, too.'

He says that the blood tests came back negative, my heart is fine, and the CT scan did not show any problems with my brain. He turns to his colleague: 'It's not TGA.' He doesn't realise that I know what this is: transient global amnesia. A brief episode of memory loss, cause unknown. I'm disappointed; it would be an interesting clinical experience to have. Thinking it might be useful, I tell him that my psychiatrist rang and thought I might have had a psychogenic fugue. He looks relieved to hear this suggestion. I mention Doctor Banister's idea that I could go to the Seaview clinic. The specialist says he will request a review by a hospital psychiatrist in case Doctor Banister doesn't get in to see me. He'll order some new blood tests and a urine test. He wants me to stay for another night so that they can monitor me, and to give him time to talk with Doctor Banister. After this, if he's satisfied, I can go to Seaview, as long as Anna takes me.

EVERYONE HAS GONE. It's a relief; without people asking me things, I slip back into a river of peace. But I try to remember to look at clocks, to keep track of time; it slips by quickly when I don't.

For the first time, I notice my mobile phone on the bedside table. If I turn it on, there will be messages, and people might want things of me. I realise how little I've thought about the troubles Anna and I are facing outside these walls. I'm not going to turn it on, for now. Instead I'd like to do something active. There is a library in the hospital somewhere, which I went to when Anna was in the early stages of labour several years ago. It has medical and psychology journals I don't usually get to see. I'll go look for it.

I put on my shoes and walk down the corridor. Each doorway I go through feels new and vibrant, like I'm a tourist in a foreign city. I follow a direction on an overhead sign, walk a short distance, and then can't remember what the sign said, or the direction the arrow was pointing in. The more I concentrate, the more my brain hurts. I realise I'm lost. Well, I'll just follow my nose.

After a time I see three of the hospital staff walking along in front of me. They are chatting and laughing, having a good time. I like their energy, so I follow them. We end up in a canteen and they sit down. It occurs to me that a coffee would be good.

As I look around, it strikes me that everyone here is hospital staff. Almost all have lanyards with photo ID tags hanging around their necks. They favour the booths; the large open area near the windows, with tables and chairs, is sparsely populated. That's where I'll sit. I stand in line to order my coffee, trying to look like I do this all the time. I'm not sure I'm meant to be here.

I sip the coffee by a window. I don't think it's very good, but I enjoy it. I watch the staff. They sit in their groups of colour: the blues, the turquoises, the whites. Some are also in regular clothes. They laugh and throw their arms around, telling stories over their sandwiches and hot food. It's like a party; they're more alive here than in the wards and corridors. Suddenly I have that feeling again: I'm not sure this is real. It's a little unsettling, now. But I'll act as if it is real, to be on the safe side.

After a while, I realise I should head back for lunch. Before I left my room, I wrote down the letter and number of my ward. On my way back, the signs are easier to follow, and by asking staff for directions once or twice, I find my way 'home'.

A nurse tells me I can be discharged that night, once they have the paperwork done. Doctor Banister hasn't been in to see me yet; I wonder if he will.

No other doctors come by that afternoon.

In the evening, Anna and our youngest daughter, Amelia, turn up. It's lovely to see them. Amelia, who is eight, gives me her bashful smile. She's keen to check out my bed and drink my milk from the little blue containers.

I'm puzzled by how quickly yesterday went. Anna says we arrived at the hospital about eight in the morning and she left at four in the afternoon. She came back around six, after the hospital had called and told her that I was trying to leave. She brought with her a change of clothes and a toothbrush. Yet I have no memory of her coming back. I tell her that yesterday seemed only an hour long.

Amelia and Anna are keen to watch the semifinal of a reality cooking show our family's been following. We snuggle into my bed, propping ourselves up with pillows, and look at the television hanging from the ceiling.

'I've worked out what today is,' I say, pleased with myself. Earlier I'd seen a newspaper lying around and caught sight of the date, and I'd been rehearsing the information ever since.

'What?' Anna says.

'It's your birthday, isn't it?'

She nods.

'Happy birthday, darling. We'll do something when I get out.'

She smiles faintly.

IT'S DARK WHEN we arrive home. The house is quiet and cool. Amelia goes straight to bed. The other two kids are away. With cups of hot Ecco, Anna and I sit down at the dining-room table.

'What happened yesterday?' I ask her.

'I got up in the morning, and you were wandering around the house. You had your business jacket on — that's the first thing I thought strange. And you asked, in a sort of dreamy monotone, "Anna, what am I supposed to be doing?" I said, "You're taking Emma and her friend Tina to camp after breakfast." A few minutes later, you asked me the same question. You were white and your skin was icy. I sat you down at the table with a heat pack around your neck.'

'I don't know if it was a dream,' I say, 'but were we in the Tarago and I … vomited? Does that make sense?'

'Yes. I wanted to get you to hospital straightaway. On the road to Lismore, you let down the window and almost threw yourself out while you vomited. We were doing one hundred kilometres an hour. I grabbed your shirt. I think you would've ended up on the road if I hadn't.'

I'm amazed.

IT'S LATE. THE only sound in the house is Anna as she goes about doing things. I settle into bed, into the silence, with the darkness closing around me. My body lets go, muscle by muscle, and sinks into the mattress.

Then, the pieces of the puzzle begin to join and a picture emerges: I've finally lost it. I've had a mental breakdown.

BEFORE

1

JUST OVER THREE years earlier, in May 2006, I had turned up to the local outdoor swimming pool on a regular Monday. I came after work each Monday and Wednesday, for the adults' swim squad.

The complex, comprising a fifty-metre pool and a toddler pool, plus a spectator stand, picnic tables, and a lawn, sat on the land's edge, with only a strip of car park between it and a seawall of rocks, which spilled onto the sand. A salmon-coloured sunset had already begun. At this time of year, it would grow in size and intensity during the swimming session, and arch over us so that we could look up at it — as we always did, grateful to live in such a beautiful place.

We were a group of varying swimming ability, having in common the desire to stay fit: a fifty–fifty mix of men and women, most over thirty years old; teachers, health professionals, businesspeople, retirees. If we didn't do this together, the invisible string that pulled us to these sessions would be gone, and we'd lapse into our individual, lackadaisical swimming efforts.

The session began with warm-up laps. After that, we stood in the shallow end, bantering with our coach, our faces turned up to him — trying to delay his next set of instructions, to give our bodies a break and allow our breathing to ease. We still had the hard work ahead of us. The squad lasted an hour, and we regularly did 2.5 kilometres, pushed to make times and distances beyond anything we'd achieve on our own. I liked this. And I liked being told what to do, not to have to think: a contrast to my workday.

Twelve years earlier, at the age of thirty-six, I had started in private practice as a clinical psychologist in Sydney. It was a stark difference from being employed: no morning and afternoon tea breaks, no collegial chats or regular meetings. I was on my own, and I had to make it work financially. Making it work meant seeing one client after another, writing report after report, all day. I also wanted to prove to myself that I could do what my training and experience had prepared me for, without a superior looking over my shoulder and without the smothering bureaucracy of the health and corrective-services departments I had worked for previously.

Private practice meant doing the unfamiliar: signing an office lease, paying for secretarial support and marketing, business networking. It also meant versatility: offering a range of psychological services, some of which drew on knowledge I had not used since my training. I put in five to six days a week getting the practice established, this schedule only easing after Anna and I had our first daughter, Ashley.

One thing I hadn't anticipated needing to learn was how to change mental gears: to go from a work mindset to being an ordinary bloke, a husband and a father. When I walked through the door each evening, I was still mentally and emotionally with my clients. But eventually I worked out a routine: after calling out 'I'm home', I'd change out of my suit and tie into casual clothes, and lie on the bed with my eyes closed for twenty minutes. After this, I was able to walk into the domestic reaches of our house as a husband and a father, the mental-health professional's uniform left hanging in the wardrobe.

After Ashley was born, we settled in a small country town fifteen minutes' drive to the beach. It was two hours from Anna's family: close but not too close. We had sought a less congested life, a place near the coast with a strong sense of community — somewhere safe to raise our daughter and the other children we hoped would come along. We were both confident that we could make a go of it. I completed locums with the local-area health service before deciding to set up again in private practice. But this time, I did it with two colleagues; I didn't want to be on my own.

So, as I swam in my lane that Monday evening, feeling the draw of the current from the swimmer before me, I could sense my mind unshackling from the mental fetters of the day. The physicality of swimming drew my concentration to each stroke, each breath, and to becoming synched with my body's rhythm.

After the session was over, we shared a few jokes to the background rhythms of the djembe drummers on the seawall before splintering off. I walked back to my car with Ian, my work colleague, catching up on news before we each headed home.

As I drove, I had hunger in my belly. I was fatigued but buzzing, my mind unclenched. I was ready for home.

THE NEXT MORNING, the routine began again. I was running late. Before I left the house, I pulled out the client files I would need that day. I was due to see an ex-nurse who suffered from chronic pain, sustained after an injury at her hospital. I liked working with her because managing chronic pain required addressing the mind and the body; it was an area in which relaxation therapy and hypnotherapy, methods I enjoyed using, were helpful.

I'd also taken up neuropsychological testing again, having done little since my practice in Sydney. I liked the intellectual challenge of these complex assessments. They required less emotional involvement — something, I'd noticed recently, I was drawing away from. Today I would need to read through the extensive reports for a young man who had suffered a brain injury following a high-speed car accident, in preparation for testing him on Wednesday. I'd also be finalising a report on a social worker; she had been stabbed by a mother whose children had been taken away from her. After several unsuccessful attempts to return to work, she was now seeking financial compensation.

When I arrived at the office, there was the usual bustle in reception. Elaine, the practice manager, was presiding over the arrival of clients and patients in her coaxing and cheerful way. In the tearoom, the kettle had already boiled for the day's first cups of tea. This was where the three partners in our practice — Ian, general psychiatrist; Peter, child and adolescent psychologist; and me, clinical and forensic psychologist — gathered each morning.

Tuesday was Ian's day for seeing patients for medication and psychiatric review. His turnover was high; he would see individuals for fifteen to thirty minutes for most of the day, generating by far the most coming-and-going among the three of us. His patients ranged from the manifestly normal to the eccentric, and, sometimes, the definitely odd.

Pete's office was next to mine. He worked with troubled adolescents and their foster carers. Every now and then I would hear muffled voices arguing, and Pete, in his low, calm tones, soothing them.

Although it was sunny, with the cool of the autumn nights it was cold inside my consulting room. Otherwise, everything was as usual: two green Ikea tub chairs; a glass-topped coffee table with a box of tissues and a Balinese statue; the grubby whiteboard on the wall to my left; and the writing desk and filing cabinet below the window. Our neighbouring property was a nursing home. Today, the pleading, high-pitched cry of 'Nurse, nurse' — it was only ever this word — had not yet started up. It came from what I imagined was a female resident with dementia, her shrunken world just metres away. Her cries usually reached a crescendo until finally (I surmised) she got a response, and all went quiet.

I typically saw a mix of psychotherapy clients, for one-hour sessions, and individuals for forensic assessments, which took ninety minutes. My first appointment this morning was with a woman in her late thirties.

Rachel had a hangdog look: long black hair without sheen hung by her face, and her eyes seemed to say, 'I don't want to be here, doing this.' She was applying for victim's compensation, and had most likely told her story many times to others. I'd read through an extensive file of documents in preparation for the assessment.

At first she was reluctant to answer my questions, but once she got going, she couldn't seem to stop. She told me how, as a child, she'd been minded by a trusted member of her small seaside community one day a week, and each time he had sexually abused her. She had accepted it as normal. This went on for years.

As she spoke, I imagined a blond brick building and saw a genial family man in his forties, with a ruddy face and a slight paunch. He was a friend of her uncle's, with whom she lived, her mother having deserted her and her father dead. Rachel later learnt that her uncle was aware of the abuse and condoned it.

I noticed sensations of disgust germinating in my body.

She said that she was not allowed to play with other children. And although her schoolteachers had suspicions that something was not right at home, they took no action.

Was that a pain in my heart? No, I must have been imagining it.

She told me that she would run away from home, returning when she became hungry. The punishments — emotional and physical — administered by her uncle became progressively more gruesome. As her story became more graphic, I was drawn into her world, as if I was there with her, watching in horror. The stabs of pain in my heart became more insistent; I wasn't imagining them. I didn't think it was a heart attack but something else, like an emotional wounding; her words were playing with my insides like a bull tossing a matador. 'Stop, stop!' I wanted to shout. But I couldn't stop her; she'd been let down by too many people in the past.

By now, thirty minutes into our interview, she was enlivened, describing one incident after another; they burst forth like spray from a water main. She described more heartbreaking incidents that occurred in her teenage years. Bad luck seemed to follow her wherever she ended up.

It was two and a half hours before we were done and she left. I had enough for my report; I wouldn't need to see her again. As she walked out the door, I wondered how she would fare. I got the impression that she didn't realise how heavily her past weighed upon her.

That night, I dreamt of my three daughters going through the ordeals that Rachel had faced, and imagined other bad things happening to them. I woke in fright and sat up in bed. *It's not really happening*, I told myself. *You're at home. It's 2006. The girls are safe in bed. You're dreaming it.*

Was Rachel's story the worst case of childhood abuse I had come across? Maybe. But I'd heard so many bad stories over twenty years. How could I say which was the worst?

Something hadn't been right with me since the summer holidays in January. I hadn't bounced back after Christmas with my usual enthusiasm for work. Now I wondered if my physical response to Rachel's story was telling me something.

MY REGULAR YOGA teacher, Flo, was hosting a weeklong yoga retreat in June. I decided to go. If I took some time out, I might find some answers to my malaise.

The retreat was held in an elevated wooden house nestled in rainforest, with an airy yoga space. After the first morning's yoga session, we all sat in a circle on the floor and spoke about why we had come. The themes were similar: stress, worry, grief, wanting to slow down, physical illness.

Flo took us through a long hatha yoga session, after which we had two hours' free time over lunch. In the afternoon, she guided us in a relaxation exercise that took us through every part of the body. This was yoga nidra, she told us. It was followed by a shorter hatha yoga session. In the evening, she guided us in a sitting meditation.

This routine continued, and as the days progressed, I felt a sense of lightness, the kind I hadn't had in a long time, as if a heavy blanket had been lifted off me. My old self began to return as I bantered with the others. The daily routine calmed me and brought my mind's eye into my body, where I saw soreness and fatigue. When images of clients inserted themselves into my mind, I noticed how the calm scattered.

On the fourth morning, I awoke with a clear thought: *Work is making me sick.* I knew that I had to stop. I needed to take a long break: recuperate, refresh, exercise, sleep — do all the things I advised my clients to do. I told the morning circle of my resolve and got nods of approval.

On the last night of the retreat, I was woken by severe cramps. Hoping to walk the pain away, I got up and went along the gravel road, which was illuminated by moonlight. I had been getting these pains in my abdomen intermittently for almost a year, and they seemed to be getting worse, and more frequent. My GP had conducted blood tests, but they had come back normal. I'd consulted a gastroenterologist, but his investigations had revealed nothing. I'd tried to work out what triggered the cramps — certain foods or activities, alcohol — but nothing was consistent. I was healthy, medically speaking, and yet my body was saying the opposite.

Finally, I woke one of the women and asked if she had medication for stomach cramps. She did, and with the early glow of the morning showing, I got some rest.

Yes, I had to stop; something was not right.

After the retreat, I began to refer on new clients. I made December the deadline to close my practice for six months.

2

THE NEXT MONTH, June, we took a family holiday amid the mountains in Lamington National Park, on the Queensland border — our favourite park. On the third day, after a long walk, we washed away the mud, went into the library at the guesthouse, and sat by an open fire, playing games and drinking tea, before dinner.

Then my phone rang. It was my sister. 'Dad's been taken to hospital. The nurse came and found him in bed, unable to get up.'

Dad hadn't called us the previous Sunday evening, as he usually did. I hadn't worried: sometimes he didn't 'get around to it' because of the time it took him to complete his daily routine, or because he'd been to a meeting for some political cause he was involved with. These calls had become exasperating, what with his increasing hearing loss. We'd shout down the line, but he couldn't decipher most of what we said. Yet just hearing our voices was reassuring for him, I thought.

'Don't cut short your holiday,' my sister said. 'I'll let you know if you need to come down.'

This was a relief. Still, I felt uneasy. Dad didn't give much away about his health.

By the time we got home two days later, Dad had undergone medical tests. 'You should come quickly. The doctors don't think he's got long to live,' my sister told me.

As I flew to Sydney, I thought of how Dad's body had, over time, shrunk around him like a prune. As a young man he had been five foot ten, but now I, at five foot eight, felt tall next to him. His knees had progressively glued up with rheumatism, becoming swollen, giving him a straight-legged gait. With two walking sticks and his body bent at the hips, he looked like a giant four-legged insect as he perambulated out of his front gate. Yet for ninety-one, his face was ridiculously boyish. Blackberry lips framed a mouth that was always ready to smile, and his olive skin was remarkably free from sun damage. The wrinkles crowding his face pointed to the warmth of his eyes. He retained a good head of grey-streaked black hair, with two wavy inlets of baldness on his forehead.

I arrived at the massive metropolitan hospital, which had opened when I was in my twenties. Now it looked weary, as if put to more use than was promised at the opening. I walked through the entrance into a foyer. The name and number of Dad's ward was on a piece of paper in my hand. I turned right into a wide corridor, which, like a train line, conveyed its passengers to the inner quarters of this medical city. The colour had gone from the carpet, except at the very edges, where it met the walls. There, azure could be seen: the pattern must have resembled a sparkling sea once.

I threaded my way through couples and families, some of them almost festive. On my left were a newsagency, a bank, and a hot-food outlet that emitted warm, fatty smells. A stall burst with tight bunches of flowers and metallic, heart-shaped balloons that announced 'It's a boy!' or 'It's a girl!'

Then the corridor narrowed. The lighting became subdued, as in a church; I had to be getting close. The carpet went, replaced by sanitised linoleum that curved up the walls, taking away the corners. Trolleys, wheelchairs, and shiny implements grew in number. I passed a tray of bedpans and urinal bottles, sitting there as if enjoying a breather before their next assignments. On my right were vending machines hoarding sweets, drinks, and chocolates — these comfort foods absurd in the face of Dad's approaching death.

Most of those walking by me were uniformed staff, feet tapping as they moved quickly, looking straight ahead, their minds perhaps curling around decisions they had made or would make. There was an eruption of laughter as I passed by a nurses' station, where nurses and doctors had congregated under a halo of lights. I was in an alien world, and my father was dying.

'Dad,' I said, as I entered his room. The top half of his bed was raised so that he was semi-reclined. He was the only one in this windowless space. There was no softness here: no jumper carelessly draped over the back of a chair, no picture crooked on the wall, no newspaper half-pushed into the top drawer. The starched bedsheet bunched like a Himalayan range around his chest. A thin cotton blanket covered his legs. The only colours were white, the gleaming grey of steel, and cream. It was an antiseptic desert.

A cable connected Dad to a rectangular box hanging off a pole. It flashed green lines that reflected some aspect of his wellbeing. His eyelids were half-closed, his mouth had sagged, and his head appeared oversized — he'd lost weight.

At the sound of my voice his eyes opened lazily, and then he gave a half-smile. 'Ah, Dave,' he said in a weak, otherworldly voice. I went over, clasped his hand, and pressed my forehead against the side of his head. He looked confused rather than scared. We stayed like this for a while, and he seemed to hum with pleasure.

I had been trained in grief counselling many years earlier, before I'd had much to mourn. We'd learnt that grief was a natural, individual process and that the most helpful thing was to avoid any predetermined ideas of how those who were grieving or dying should act. I'd seen differing reactions before: relatives who felt compelled to cheer up the dying one, offering comments such as, 'Well, it's a nice room, isn't it?' or 'You're looking more perky today' — trying to keep their anxiety, their uncomfortableness, from spilling over. Dad would have seen through such behaviour even if I could have managed it, and then he'd want to comfort me.

For the next seven days I came every day, staying at Dad's house nearby. We often sat in a kind of stationary silence. Now and then, we talked. I was moved when Dad told me how fortunate he was to have me and my siblings, our partners, and his grandchildren. There was always an update on the environment petition he had been working on for years, and he let me know which politicians or activists had responded to his last round of letters. 'How are Anna and the children? How's my house?' he'd ask.

I didn't say that Anna was at the ready to fly down with the kids as soon as I gave the word; that the house was a shambles, its corridors and rooms filled with the collected years of hoarded junk. Most of the community workers my siblings and I had tried to get to assist him had refused to come into the house, citing health and safety concerns. Once, we had organised a private meal service, scheduled to deliver a hot meal each day. After two deliveries, he told them that the meals were very nice but he could cook his own, thank you.

I told him that the house was fine. Anna and the girls were well and thinking of him.

Each day, when it felt like the sludge of my grief was going to suffocate me and the clinical feel of the room became too much, I said, 'I'll go now, Dad.' He would simply smile and thank me for coming.

DAD HAD BEEN admitted with angina. The doctors said that he was also weak from inadequate nutrition. A few days into his stay, my sister and two brothers had joined me for a meeting with the senior physician. 'Your father has a very weak heart,' he said. 'It's worn out and could give up on him at any moment. There is also something else making him sick; we don't know what it is yet. If he ever recovers from this, he'll need high-dependence care.' He explained that this meant twenty-four-hour medical care: a nursing home. 'Your father will never return home,' was his summary statement.

About a week later, doctors found that Dad had a cyst, and it was this that had been causing him increasing pain. They operated successfully. For several weeks it was touch-and-go, and I returned home. Eventually Dad regained his colour, and we began to think about moving him to a nursing home.

But then Dad stopped eating. This became the new health crisis. I travelled back to the hospital, where a junior doctor spoke with me. Her inexperience was palpable, and I wondered how she would respond to the years of human suffering that lay ahead of her.

'We think your father is depressed,' she said.

A decision was made to give Dad antidepressants.

But for weeks there was no significant improvement: his weak heart and lack of eating stalled his recovery.

After a month, and to our great surprise, the hospital told us he was to be discharged. How could this be? When I looked at his sunken chest, his birdlike shoulders, and his melon-sized head, I couldn't reconcile him with the man of my youth. The father who had showed me how to use tools and make things, who could push a wheelbarrow and shovel dirt all day in the garden, his rolled sleeves revealing bulging veins. The father who, when we were very small, carried me on his shoulders, giving me instant height, and played 'horsies', as three of us at a time piled onto his back while he laboured on hands and knees, moving forward jerkily.

I understood that the hospital could not let Dad take up a bed while he decided if he was going to eat. We were given two weeks to find a nursing home of our choice; otherwise, he would be deposited into the next available bed. My brother and sister found a place in an Anglican establishment. The staff appeared caring and the place looked well run, they said.

A few days after he moved in, I visited him. His room was large, and he shared it with three elderly men. The one opposite him had no legs and lay in his bed unmoving, now and then vocalising incomprehensible sounds. Dad said that he could joke with the man diagonally opposite him, but, lowering his head, murmured that the next man along was a 'sourpuss' and wanted the air-conditioning up too high all the time. By Dad's bed was a large window with a view of an expansive, tree-lined park where mothers pushed prams, and joggers, cyclists, and dogs all paraded.

Now that he couldn't go out, the world had come to him. I arranged for him to have 'pocket money' so that he could buy the little extras he wanted. The tables had turned.

Dad started eating, began to write his letters again, and charmed the staff. I went back to my normal life, uneasy about how long this period of equanimity would last.

THE AFTERNOON I returned home, I made my way to the beach: I needed to settle, and to regain perspective. I walked along the foreshore. European backpackers played hacky sack in small groups, while others knocked around a soccer ball, creating a carnival atmosphere. The surf lifted my mood.

But the bay also offered its own moods. The beach was exposed to the northerly wind, which was at its most blustery in spring. It should have brought warmth from the tropics, but at this time of year, when the sea was at its coldest, it whipped off the water, turning it cold. It bullied the ocean swell, which hissed up in green walls of defiance, foam fuming from its peaks. It screeched onto the beach, throwing sand into the faces of the few determined souls that strode along, clothes flapping, their heads lowered in submission. It swept over the fence of the open-air swimming pool, sucking the warmth from the water. The children in the after-school swimming squads, when finally released from their chlorinated chamber, stood around the pool's edge like shivering penguins, their towels draped around their shoulders.

The seabirds that scouted from the sky, looking for schools of baitfish, didn't like this wind: they couldn't spot prey in these conditions. But the majestic Norfolk Island pines between the swimming pool and the surf club, with their crinkly trunks and symmetrical branches, danced. Energised, their limbs resonated with the bluster.

After spring waned, the northerly reappeared at brief intervals in summer, this time as a hot, sticky wind. It pushed Portuguese man o' war, or bluebottles, onto the beach. Sometimes swimmers would notice these opaque, air-filled bubbles floating on the water, with long, iridescent threads of poison hanging below, and most knew to avoid them. But when stung — and it often happened — adults groaned and washed the invisible barbs off their red welts under the freshwater showers; kids screamed in fright as much as in pain, and rushed towards their parents.

For most of the year, the giant prism of the cape at the southernmost point of the bay headed off the southerly ocean swells. But in winter, the massive swells rumbled around the corner, growling into the bay with big-muscled bravado.

The surfers loved them.

The mighty tubes, spray spitting from their glassy peaks, became impregnated with black, stick-like figures on surfboards, and then collapsed like thunder. Tailfins shot into the air while bodies flayed and star-jumped into the water.

Somehow, the shifting moods of the bay always soothed or elevated my own. I never understood quite how this worked, but it seemed that in the bay's enormity, with the crescent of green that clung to its foreshore, I was always accommodated.

MY GRADUAL WITHDRAWAL from work was a relief. By that August, I was only coming into my practice once a month, to see a handful of long-term clients. But the nights were penetrated by bad dreams, and it took me up to an hour to get to sleep. The more I let go of work, the more I unravelled.

I brought my old cricket bat into the bedroom, resting it by the bed, within arm's reach. Although I'd never heard of anything bad happening in our neighbourhood, I feared that we might experience a home invasion. I especially feared for the safety of the girls and Anna. Anna said that I was overreacting.

The kids' noise and demands had become irritating. It was okay when they didn't squabble, but they were ten, seven, and five — squabbles were inevitable. Sometimes I wished they'd go away. Anna prodded me to do things, and I wasn't sure if she was being bossy or I was being slack. My energy for gardening disappeared — the weeds and wilting plants taunted me. I felt uncomfortable in crowds and didn't like confined spaces. Other than my swimming, most things required too much effort. On weekends I sat for long periods on the verandah, reading the newspaper. I was drawn to articles about murders, disasters, and neglect of children, even as I was repulsed by the stories.

Our family attended Nippers, a surf-lifesaving program for children, on Sunday mornings. But I was frustrated when the girls held back, fearful of going in the waves. I didn't remember being afraid of the waves when I was young.

What was happening to me? I was prickly, yet as fragile as porcelain. What I perceived as aggressive words from others frightened me. Watching the news, I had begun to cry at pictures showing human or animal distress, and even at sentimental, good-news stories. I had an idea I'd seen too much human suffering, especially trauma. But I wasn't sure if that was the whole story. Images of Dad's near-death and what might happen next sloshed around in my mind.

I knew that I needed help, but from whom? I was the helper — the one others always relied upon to know what to do. I'd talked to my colleagues, but how much could I burden them? I needed someone else; I couldn't do this by myself anymore.

I thought of the psychologists in our area. There were not many with more experience than me. I called Wayne, a clinical psychologist. I knew he had done his PhD on psychological trauma in Vietnam War veterans, and we'd spoken a couple of times, over the phone, about trauma cases. He was the best I could find. I made an appointment for September.

3

WAYNE'S OFFICE WAS on the first floor of an old building in a nearby town. As I walked up the staircase, each step felt like a deepening admission of my crushed sense of invincibility. At the top, there was no receptionist. I waited in a poky room, alone, sitting on one of the worn chairs. A radio bleated from the corner.

Out of a door came a tall, broad-shouldered man in his late fifties, dressed like a farmer in town clothes. He had florid hair — a style that was a throwback to the seventies — a fleshy face, and a generous mouth. 'David?' he said, looking at me.

'Yes.'

'G'day. Wayne.'

He invited me in. Like mine, his consulting room was unadorned; there were no personal items, except for framed degrees and certificates. But it was light-filled, with a big open window. The noises of cars, conversation, and a busker's banjo floated up from the street below.

'Have a seat over there.' He pointed to a two-seater sofa. Then he sat down on a swivel chair by the desk, below the window.

He turned to face me, brought his chair closer, and crossed his legs, resting a clipboard on his lap. 'What can I do to help — what brings you here?' he said. There was warmth in his voice, although his face was impassive.

I leant forward, resting my elbows on my knees. I noticed the chunky soles of his shoes. 'I … I don't think I can cope anymore,' I told the carpet. I could hear a quaver in my voice. 'It's hard to describe. I'm — I don't want to go into the office anymore. I went on a yoga retreat and I felt freed from everything. Umm, I realised how sick I've been. I've had this bad back, and … I've been getting pains in my stomach; the pains come and go.'

Then I told him about Rachel and the response I'd had.

'Do you still think of her?'

'I think of — I think of how bad her life was as a child, how cruel her uncle was, how a society can let that happen. I think of my girls. I couldn't handle it if they went through something like that.'

'Are you thinking about this in words or are you imagining it in scenes?'

'I see the girls in horrible situations, maybe … being abducted or raped or …'

'That's terrible!'

I lifted my gaze. Wayne was leaning back, eyes wide in alarm. 'I'm a father myself,' he said. His reaction surprised me: what I was saying must be bad. His eyebrows knitted together. 'How have you managed with this? Have you had anyone you can speak to about it?'

'Yes, that's it. I don't have anyone I can really talk to. Anna's sympathetic, but she can't understand. She wants me to tell her about the bad things I've heard, but I don't want to traumatise her with it, and it's confidential, anyway. Ian, my mate from work, we talk. But he's caught up in family and work. I have a peer supervision group; they know I've been having a hard time. But we only meet once a month. I've really come out of desperation …

I can't do it on my own.' I was talking to the carpet again.

'Are you thinking you will continue with your practice?'

'I think I can get back; I've given myself six months off. We have some savings, so it's not desperate.'

Wayne asked me to tell him about my sleep. I told him that I was going to bed early; I got very tired but it took a long time to fall asleep. When I did, I was troubled by nightmares, mostly of things happening to the family. 'There's a crime show we watch on Fridays. There was a scene where the owner of a Chinese restaurant is stood over by these heavies who want money. When he says he doesn't have it, they put his hand into a deep-fryer of boiling oil and hold it there. I know these are shows, but they seem real to me. The scenes are stuck in my brain; I can't get them out. It's getting worse. I break down in tears watching the news; I can't stand bad things happening to people. I feel like a broken-down machine. I've failed.'

Wayne leant forward. 'How have you failed, who have you failed, what have you failed at?'

'It feels like I've let my mother down. She was a psychiatrist and went through her whole career without this happening to her. I've got other friends who are psychologists who seem to be doing okay. So why can't I do it anymore? I've got all this training and experience. I'm at the height of my career and I can't help anyone.'

Wayne asked if my mother was proud of my work as a psychologist.

'Anna told me that Mum used to listen to me on the radio when I was a regular guest. I don't think she minded what I did, as long as I was happy doing it. I know that she loved me, but she wasn't an outwardly affectionate sort of person. One day, in my early twenties, I came home for a visit and gave her a hug; she liked it, and after that we always hugged. But Dad was the more affectionate one.'

I told him about Dad. I also said that on Sundays I swam in

a group, the Stingrays, and we caught up for coffee afterwards. This had helped.

'Do you ever think of doing away with yourself?' Wayne asked.

'Yeah, I've had thoughts about that. It scares me.'

'How serious?'

He's checking my level of suicidal intention: do I have a thought-through plan, have I made any attempts.

Suicidal urges had sneaked up on me, like unwanted acquaintances tapping me on the shoulder, wresting my attention. I hadn't told anyone before. I had pictured myself jumping off the cape, a huge headland near home, onto the boulders below, and imagined my lifeless body being raised by the sea's swell and carried away. I had imagined slipping quietly into the water at the edge of the beach — just another swimmer in the early evening — and propelling myself further and further out into the growing darkness, breathing through the fear that would surely come, until I tired, or something took me under.

But, like reining in a bolting horse, I'd managed to pull these thoughts up short. I'd considered those left behind. I couldn't let the girls grow up without a father — who would protect them? I knew from my work that suicide and the death of a child were the most heart-rending of deaths for those left behind, creating a gut-twisting cocktail of incomprehension, sadness, guilt, and anger.

'Do you agree to tell me if you really think you would?'

'Yes,' I said, looking at him squarely. I knew how necessary it was for a clinician to hear this. I didn't mind reassuring him: I hadn't believed that I would really go through with it.

'You can ring me at any time.'

He wanted to know when I got the uninvited, disturbing thoughts or images that changed my mood.

'If I read about something in the newspaper — a murder, or something bad happening to a child — that sets me off.'

'What about just out of the blue?'

'Yes, yes … I can't recognise any pattern to it often.'

'When some people have a panic attack, they get a sense of dread that something bad is about to happen, and it can be very physical.' He swept his arm through the air as if including all possibilities.

'Yes, I get that. It feels like I'm under attack: my body tightens, shakes. But I don't know what the threat is, or where it is. The world's an unfair place; the weak are trodden on. Our society is dislocated, falling apart.'

He asked how I was getting on with Anna and the children, about other physical symptoms, about my medical and family history, and then about drinking. I said that I'd been having two to three glasses of wine a night — more than I used to. I tried to have two alcohol-free days a week. But it was getting harder to take my mind off drinking.

Towards the end of the session, Wayne said, 'I think you've got a post-traumatic stress reaction and the depression that goes with it. Really, your battery is very flat. You've not had healthy sleep now for so long. Psychologists are witness to an enormous amount of suffering. We do our work in a closed room and we can't share our work experiences like those in most other occupations can.'

So I hadn't been imagining it. A diagnosis of post-traumatic stress disorder was heavy stuff; I really was sick.

'Stop watching the news,' he said. 'Nowadays, you're going to see genuine grief, multiple incidents of violence, in one half-hour. It's necessary to start protecting your imagination. Exercise. Simplify your life.'

He wanted me to write down my dreams. I was to ask myself a deliberately open-ended question before I went to sleep: *Would it be okay to remember a dream about the problem I am facing or about its solution?*

WAYNE AND I caught up again in a fortnight. I described one of my recent dreams while he listened attentively. 'I'm driving a van. The three children are in the back seat and Anna's in the front. We're going down a winding road towards a bridge over a river. It's late in the day and the light is failing. The van crashes through the railings and plunges into the water. The girls scream. I look across at Anna's ashen face. We are sinking. Do I wind down the windows so we can swim out, or is it best to keep everything shut so we still have a pocket of air? Movie images of people trapped in ships, boats, cars flash through my mind: what did *they* do?

'The water is squirting through the gaps in the van; it's murky and getting dark. We can still breathe, but it's hard to see the children. "Anna, grab the one nearest to you!" I shout. "Push yourself out the door!" There's two girls left — I can only help one. I grab Amelia and yell to Ashley to follow me. I push against the door … The water pressure is immense. I'm not going to make it … Then I wake up hot and shaking; I think I've just died.'

I told Wayne that I remembered the nightmare protocol, a series of questions to help orientate the individual that I used to give to clients: *Where am I? What's in the room? What time is it? Am I safe?*

Wayne wanted to analyse the dream. 'What is the dictionary definition of a van?'

'It's a form of transport.'

'What emotional association does a van have for you? What could it represent?'

'The van is the family and me together as a unit.'

'And what does the road represent?'

'It's a pathway, a way of getting from A to B. The road is narrow and we are stuck, hemmed in, trapped.'

'What does a bridge refer to?'

'It's a means for getting over things, like a river or a canyon. It can collapse.'

'And water?'

'It's a liquid; it can be life-saving or it can take away life.'

Wayne offered an interpretation. 'The dreamer is on a journey, the journey the family is undertaking. The family is being constricted — trapped and suffocated by what's happening around them, what's being done to them. The family is being drowned in the heaviness of sorrow and the weight that you feel. You and they are stuck in a place in which you can see no way out. At present, the dreamer cannot clearly see any means of survival for the family. He feels that he has lost control and is unable to save his family.'

What he said felt right. I had lost control. We were under threat and I could not save my family. But why was I traumatised when I had never been physically harmed?

'It's the threat of harm, and feeling what others are feeling, when they describe what they've been through, that has injured you,' he said. Each awful story I'd been told had been a nick in my psychological armour. Rachel's story was one too many; my armour had shattered. Wayne said that the nightmares were re-traumatising me; the terror I had experienced was real. My imagination was a gift and a curse: it let me put myself into others' shoes, but it also concocted vivid images that damaged me. My capacity for empathy had become a poison.

Wayne told me that he had stopped working intensively with war veterans when their horrific stories had become too much. Now he took fewer appointments and continued to see only a handful of long-term clients, and instead presented more workshops and provided clinical supervision.

I felt that he understood. But I couldn't let go of the sense of failure so easily. I thought I'd let everyone down: my family, my profession, my community, the taxpayers who had helped to fund my nine years of training. I was a sought-after expert; now I was useless. No, even worse — I was a liability.

Wayne said that I was in a transition phase. My time at the coalface of clinical psychology was probably over, but I could draw on this experience in the next phase. He told me that all psychologists at some time in their career confronted this dilemma: how to maintain empathy for their clients without taking on too much of their suffering.

IN THE FOLLOWING sessions, I reported more dreams — some that involved the family, some in which I tried to save others, some in which I died, some in which I escaped. Wayne, using a cognitive behaviour therapy approach, challenged my thoughts of failure and sense of worthlessness.

In one of my dreams, I was riding a bicycle over the tops of prison cells, which had a wire-mesh roof so that I could see inmates below running and jumping, trying to catch me. I managed to escape. Wayne saw this as positive. He asked me to think of my nightmares as healing dreams: my unconscious mind processing the bad memories, clearing the decks.

And slowly, like an ocean liner being turned in the harbour, I began to change tack. Even though I still *felt* like a failure, I sensed that I could have a future that drew upon my past. I didn't have to blame myself: bad things happened to good people through no fault of their own. My intentions had been good. I was doing my best.

'No matter what has happened to you recently,' Wayne said, 'that has not changed all the help you gave to hundreds of others. Your clients' lives are better for having met you. Be easier on yourself. Start thinking about the next stage of your life.'

I realised that going through pain changes you. I could never be the same person I was, but perhaps I could begin to think about becoming someone new.

4

'OPEN YOUR MOUTH wider, see, like this.' Lily demonstrated by giving an exaggerated, carefree laugh while sitting at the digital piano. I chuckled. A mental picture of a mythical wide-mouthed frog came to mind.

We were in her lounge room, a short walk from my house. Now that I was not working as much, I had the time to take singing lessons. Today Lily had been playing scales, up and down, and getting me to shape my mouth around the different vowel sounds.

She turned side on, her face looking up at me, impossibly bright for this time of the morning. As she had instructed, I was standing with my feet apart, holding a solid core, one hand on my abdomen to gauge its movement. She had placed a mirror in front of me. It was disconcerting to watch my freckled face, with a nose that always looked bent, contorting into these facial poses. 'Over-exaggerate,' she exhorted.

My early years had been spent at a small Roman Catholic convent school. One day, when I was eight, I and some other classmates were pulled in from the playground and made to stand by the school's cranky piano. In turns, we were asked to sing a song about flowers growing in a garden, while one of the nuns played. This audition led to me being selected, together with a handful of others, to attend a major eisteddfod in the city.

A few lunchtime rehearsals followed, after which several of us boys piled into Mum's station wagon, some lolling like seals in the back cabin, and set off on a ninety-minute drive to the city. It was early summer, and we took bags of cherries to munch on. By the time we arrived, I was feeling sick, although still enthusiastic about the performance.

Until it happened.

I was pushed onto the stage of the biggest hall I had been in, hundreds of faces looking up at me, the only other person with me an unfamiliar pianist. Fear grabbed me by the throat. The woman at the piano played the intro for my song, but when I was supposed to sing, nothing happened; I was still staring at all the people.

The piano stopped.

'Shall we start again?' the woman asked in a kind voice.

I nodded.

The second time around I actually made a sound, and then held the tune to the end of the song. The adjudicator awarded me a highly commended, with a comment: 'Open your mouth more and your voice will sound even better. Good work.'

I had found out I could sing, and that I quite liked it.

As a teenager, I had played in a high-school rock group and fancied myself a singer–songwriter. During university, I'd played in a semi-professional dance band. Yet while I'd had years of guitar lessons — Dad insisting that I learn classical guitar, to get a good technique — I'd never trained as a singer.

Now, who better to take lessons with than my friend Lily? We'd met in 1990 in Canberra, where I'd moved to take up a university position while finishing off my PhD research. I'd won a local amateur songwriting competition, the prize being time in a recording studio; Lily, then a voice student, helped me with the backing vocals. A few years ago, she and her family had moved to the same country town as us. Our agreement was that she would give me a lesson one morning a week in return for a bottle of aged wine from my collection. After each session, we stopped for tea.

Lily had recorded a CD with a series of scales and exercises that had become my vocal homework. In the morning, when the kids were off at school and not much else seemed worth doing, I went into the room downstairs, closed the windows — so the neighbours were less likely to hear — turned on my CD player, and sang to her recording, moulding my face and breathing as she had shown me. And my lumpen body and despondent mood responded.

Without a weekly lesson at which I was required to demonstrate progress, I would've forgone these exercises. But I couldn't disappoint Lily. When I turned up at her house, she'd call out with a singsong 'Halloo.' Her joyousness, and her commitment to willing me on, didn't let up from that moment.

In time, Lily had me singing an Italian aria, 'Amarilli, Mia Bella'. It was a love poem; I'd never sung anything like it before. It was a good song, she said, for people who were changing their vocal technique. My voice was too nasally and needed to come from the belly.

'You have to get out of your head. Feel love in your heart,' she said, placing a hand on her chest, her face expressing yearning. 'Imagine the person you're singing to.'

I wondered if the song choice was not only about vocal technique, but also Lily's ploy to get me more in touch with expressing my feelings.

After weeks of this, I said, 'Lily, singing is different from playing an instrument; it wakes up the whole body. The body is your instrument. You can't be depressed and sing at the same time.'

She smiled: this revelation wasn't new to her.

The discovery that I couldn't sing properly while slumped in despondency showed me that working the body in certain ways changed one's mood. It bypassed the head — that ruminating heavyweight in which I was so often stuck.

Lily's choice of songs had extended me, vocally and emotionally. I had discovered that I was a baritone, with a two-octave range: from a resonating low to a flute-like high. Singing awakened a joyousness that I thought had been quashed — a feeling that came with the beauty of the sound I could make.

I HAD BEEN visiting Dad intermittently over the last fourteen months, since he'd been in the nursing home. Now in a wheelchair, he used his feet to paddle through the corridors, and, like a lizard, he fossicked out the sunny spots during the day's diurnal changes. He liked the food they served and had bulked up, his sunken cheeks now a memory. He didn't ask about home anymore.

Dad enjoyed socialising with the staff and other residents, who soon learnt about The Petition. Since his retirement, he had engaged in environmental activism — in particular, in starting up petitions to promote various environmental causes. He was a ceaseless letter-writer, and in an age of digital speed, he wrote long letters in a feathery-like script. He used to collect signatures in public places, setting up a table and chair. When my siblings and I were younger, he had embarrassed us many times with the way he pounced on any new friend or visitor who came to our home, asking, 'Have you signed the petition yet?' with a cheeky grin. After the expected 'no', he'd passionately argue his case for protection, and the newcomer would duly sign. His petitioning efforts brought success, too: he became a formidable force in advancing awareness of potential dangers to the Great Barrier Reef, and his amassed signatures helped in the ultimate declaration of the area as a marine national park.

In the nursing home, the deputy director of nursing had, like a colonial governor granting parcels of land, given him a wedge of territory in her office — a drawer in her filing cabinet — where he stored his papers and correspondence, away from the cleaners' zealous clutches.

Sometimes, on my visits, I found him in his room, in the wheelchair, by the windowpane tinged with the gold of afternoon sun, asleep. His papers sliding off his lap, pen limp in his hand, chin resting on his chest, and a line of spittle hanging from parted lips. And in that sagging body I could see the boy he once was, still driving this old man. If a painter were asked to interpret the scene, he might have depicted a boy in old man's clothes, climbing up the mature eucalyptus tree outside the window, determined to explore.

Then, in mid-October, I got an unexpected call from my sister. 'Dad's not eating,' she said.

This was a federal election year. Dad had been working with a preoccupied feverishness, writing letters to all the politicians he thought would promote his views. When I went in to see him, I learnt that he had been off his food since he began his letter campaign. He drank enough of the protein milk given to him in the little cartons to keep the nurses off his back, but my siblings and I found bits of food hidden away in his drawers and under the bed. He told me that he was trying to fool the staff into believing he was eating more than he was. It was a bit like me when I was a boy, slipping the dreaded peas under the dining-room table and out of sight.

A couple of weeks later, once his last round of letters was posted out, he stopped eating altogether. I asked him why, but he didn't really answer. I wasn't sure if he was being evasive or if he was just tired by the line of questioning.

He was becoming skeletal again.

The deputy director of nursing asked my siblings and me if we wanted him to be fed by a nasal tube, in which case he would need to be transferred to the hospital. I thought of that white, antiseptic room. Dad liked it here — it had become his home. My siblings felt similarly; we agreed that we would encourage him to eat but we would not sanction forced feeding. My father was showing obstinate self-determination; that was his character, and I admired him for it.

DAD AND I had spoken about death before. I knew from these discussions that he did not fear it, and this comforted me. I did not need to reassure him about something that was a mystery to me.

There were no unspoken grievances between us. Two years earlier I had undertaken an online course taught by professor Martin Seligman, a founder of the positive-psychology movement, a relatively new field of social science that focused on increasing contentment and happiness in individuals, groups, and institutions. One of the exercises had been to write a 'gratitude letter'. The aim was to choose someone important to you, tell them the ways in which you felt grateful to them, and thank them for the positive things they had done for you. Then you met with the letter's recipient, if they were still alive, and read it out to them. Seligman told us that this had a positive and lasting effect on the letter-writer.

I had sat almost knee-to-knee with Dad by his dining-room table while I read out my one-page letter. I told him of specific incidents in which I was grateful for what he'd done, and how

I'd always felt supported and loved by him. I got choked up, but Dad's encouraging smile kept me going to the end. 'Thank you, Dave,' he'd said. 'You've been a wonderful son.' After that, the kettle went on the stove and we had a cup of tea: our family's ritual.

At the time, it had seemed irrelevant to tell my ninety-year-old father of the things that made me unhappy with him, the disappointments. As a parent, I knew how easy it was to make mistakes, even with the best intentions. What I hadn't foreseen was how this event released me from the need to say anything further to him now.

While I was in Sydney with Dad, I stayed with my old friends Peter and Ros, even though they lived on the other side of the city. I didn't want to be on my own, and they offered me their garden studio for as long as I needed it. Each day, I caught a bus and then a train to the hospital. Dad was noticeably weaker and less alert with every visit. I usually only managed a few hours with him; any more and the heaviness would've drowned me. But I knew that I was spending the last days of his life with him. It was a difficult but important time. As I sat beside Dad, I stroked his arm. I remembered how, when I was a boy, he had sat by my bedside at night, tousling my hair with parted fingers. In the glow of the lamp, he'd answer my questions from the day, before sleep's final wave rolled over me.

Four days after I began to visit Dad daily, on 7 November 2007, the deputy director of nursing said his end was close. Dad was moved into a room by himself. A new phase had begun.

My sister, my older brother, and I gathered by his bedside. Our younger brother had left work and was on a train to the hospital.

By late afternoon, Dad's breathing was terribly laboured, and we took it in turns to have time with him alone: my brothers first, and then my sister, and then me.

I rested my hand lightly on his chest. The pillows propped up his head and shoulders. His face was so gaunt, his shoulders barely there, and his skin as white as a marble bust.

He was trying to speak. I leant in close.

'Hi … huff … flew,' he said. His mouth had trouble wrapping around the words; each required a whole breath, and came out with the same explosiveness as someone puffing out birthday candles.

His hands rested on his lap. I clasped them in mine. 'What is it you want?' I said, looking into his face.

With great effort, his shrunken chest filled with air, and I heard again, 'Hi … huff … flew.'

What was he saying? My bafflement was distressing me now. 'Do you want me to call the nurse?' I glanced behind me to the doorway, hoping she might be coming around the corner. I was not going to leave him.

Then, his bony, elegant fingers took my hand tenderly. He lifted his head from the pillows, brought my hand to his lips, and kissed it twice. He let my hand down gently, his head slumped back into the pillow, and his eyes closed.

I called in my sister and brothers. Soon, a deep gurgling rose from Dad's throat, although his mouth remained open. His cheeks were hollows. He was losing his humanness.

The deputy director came in soon after, followed by a doctor. After the doctor examined my father and left, the deputy director said, 'His personality has gone now. His brain is on automatic pilot.' She left us to be with him. That's when it sunk in: he was never going to come back. Ever.

My eyes began to water. I glanced at my siblings standing around the bed, all silent, with a touch of horror in their eyes.

Dad's convulsive breathing continued, and then (why was it this particular one?) he drew his last breath. The silence was so sudden that I was not sure if his body's life force had really gone, if his breath would erupt again. But he was unmoving, and no matter how intensely I studied his face for a sign of life, I could not urge it back into existence.

That evening it struck me: Dad had been saying, 'I love you.' The same fingers that had caressed my hair as a boy had spoken again of his love. I thought about how one breath separates life from death, father from son, and I wasn't sure if I said it or thought it: *Goodbye, Dad, and thank you.*

THE DAYS THAT followed involved funeral arrangements: contacting Dad's friends and relatives; writing a eulogy. Anna and the children joined me. We located an eco-friendly coffin, made of thick recycled cardboard, covered by the image of a magnificent forest tree; Dad would have liked this. The wake after the service was a happy one. Many felt affection for Dad; he'd made a difference, and he'd left this life without regrets.

When it was over, I was keen to put Sydney behind me and get back to the bay. On my first Sunday back, Ian, Lily, and I rounded the surf club for the start of our swim, and my heart leapt. The beach looked idyllic. Before us, the aquamarine sea was laid out like a banquet, and the swell was nosing onto the beach. It was as if the sea was conscious of how magnificent it looked. The breeze was not up yet, and we walked through a stillness that somehow felt alive.

The Sunday-morning ritual had begun with the familiar sound of Ian's Subaru Forester pulling up in my driveway. He had let himself into the house. The rest of the family were asleep. I had already squeezed in a short meditation before he materialised, with his jocular 'Hello,' in shorts, sandals, and a T-shirt.

'Hi, Ian. Meet you out front?' I said, and went to the garage to collect my swim bag and towel. Then I walked along the side of the house, up the driveway, to his car, and plonked into the passenger's seat. 'Can we pick up Lily?' She sometimes called to say she'd like to join us.

'Yep.' Ian turned on the ignition, and his favourite radio station trumpeted out from the speakers.

A few minutes later, we stopped outside a blue timber house. Tall tiger grass in need of a trim blocked most of the view to the house from the road. A tradesperson's ute and a rusty Toyota Camry wagon stood in the drive. I got out and peeked through the grass to see a beach towel flopping out of an African basket that had been set upon the porch. The front door was ajar. These things meant that Lily would appear soon.

In a few minutes, Lily came out, blonde hair tousled. She'd probably only been up for a short time; usually she'd had a performance or she'd been to see live music the night before.

She came and lowered her face to my window. 'Halloo.' Then, with basket and handbag, she nestled into the back seat.

It was the familiarity of this routine that was comforting. Sometimes, when the week was an act of survival — as it had been many times in the last year — the Sunday swim was the main thing I looked forward to. And I knew that whatever miserable state I was in, my swimming buddies and the ocean would accommodate me.

This morning, sixty or so swimmers were gathered. We walked into a cloud of excited conversations, as well as hazy greetings from those still waking up. Then, at the agreed time, the group trod down the stairs and onto the sand. For twenty minutes, we strolled up the beach at an easy pace, chatting in duos or trios, small groups forming and re-forming. Goggles and caps were tucked into swimmers or dangled from hands; some carried flippers. As we walked, I looked beyond the waves to see if there was a current running. Was the surface smooth, so that

I could glide through it, or was it becoming choppy in the lifting breeze, so that the waves would slap my face and slosh into my mouth as I turned to breathe?

Once we reached the end of the beach, we all entered the water to swim back to the club: a distance of about 1.5 kilometres. The swim would take around twenty-five minutes, depending upon the strength of the current.

When waist-deep, I dove under the first few waves and began to freestyle. The water whistled and sizzled in my ears. My hands pushed through it, my elbows lifting high, shaped like the bent wings of a gannet diving for fish. Each arm pulled back in an S-shape, the hand cupping the water and thrusting it past my hip.

On a day like this, I could see the bottom of the ocean. The ripples of sand, like dunes in the Great Sandy Desert, curved and snaked, displaying the vagaries of the invisible micro-currents. Fish weaved lazily, and then bent and darted off. A cloud of sand swirled up every now and then, as if an invisible urchin was stirring the ocean floor mischievously in order to surprise the passing swimmer.

I glided over a turtle as it hovered by some waving seaweed. Disturbed by my presence, it turned and moved with a grace that belied its bulbous shape, flying like a bird, its tail flippers steering gently.

The swim was therapy, a purging of the week's psychic stains. Tensions sloughed off like discarded skin. Today, I felt it was washing away the intensity of the last fortnight.

In front of the surf club, I turned and made for the beach. At the water's edge the ground raced up towards me, as if I was an aeroplane falling out of the sky. I could make out single grains of sand and shell fragments — bone, yellow, black, brown, clear silica. I emerged out of the sea, at first unsteady on my feet, my mind clear and light. I was home, at last.

Afterwards, Ian, Lily, and I joined our other swimming friends and went for coffee. Our conversations usually moved from the superficial to the humorous to the personal and back again. Everything was up for discussion — health, our children, relationships, movies, books, and politics. Years of this routine had lubricated the ease with which we talked and joked. The others asked me about Dad's death, the funeral, and how I was feeling. Some of them had had parents die, and I asked what it had been like for them. At this time in my life, and in the aftermath of my father's death, the Sunday-morning swim was more than just a swim — it was a life rope.

5

THE LAST TWELVE months had been difficult — it seemed as if I had been hurled from one crisis to another. By early 2008, it was clear that I was not going to return to full-time work. We'd have to draw further on our financial investments to stay afloat. This, along with the strain of my mental and physical health problems, was taking a toll on Anna, and on our relationship.

I had met Anna when she was in the final year of a visual-arts degree. I was in the midst of a PhD in clinical psychology, working part-time at my university. A mutual friend, who had been trying to get us together for some time, introduced us. What first struck me about Anna was her stunning Mediterranean complexion. In appearance we were opposites: she, fine-boned, with long black hair, brown eyes, and an olive complexion; I, a sturdy physique, fair skin, blue–green eyes, and auburn hair. Her parents had separated early, and events had transpired to create a difficult upbringing for her, the type that made a child either sink or swim. Anna swam. It may have been this that gave her an easy maturity beyond her years, which I found seductive. She was twenty, and I was thirty-two.

When she'd first invited me back to her student bedsit, the main room filled by a sofa the size of a continent and a spongy double bed, she made baba ghanoush for lunch. This was a favourite of mine, one I'd thought only came from the Lebanese restaurants along Cleveland Street in Sydney. I watched as she placed a taut eggplant on a wire-gauze stand perched over a Bunsen burner. As the skin burnt, giving off a smoky pungency, she turned it until it collapsed into itself. Then she mashed the cooked flesh into a paste with tahini and lemon juice. I was impressed.

Over the coming months, we'd discovered how aligned we were. We were both comforted by nature. We camped in national parks by creeks and rivers, and lay in their cascades of water. We made love under the trees beside our tent. I marvelled at the inventive ease with which she made meals on our campfire.

After she moved into a share house, I made a vegetable garden for her. I sheet-mulched the lawn with disused student-information booklets, added manure and straw, and then planted it out. This gave her a season's worth of herbs and vegetables at her doorstep.

Without noticing at first, we started whispering our dreams to each other — dreams that included both of us. We agreed that we'd like to live in the country outside of a major city, on acreage, and grow fruit and vegetables. I would like to be near the sea. We both wanted children.

After seeing me meditate one morning on one of our camping trips, Anna said that she'd fallen in love with me.

A year or so after cooking that first batch of baba ghanoush, she looked at me and said, 'We have these big dreams, Dave, but where are we going with this? Are we really going to do these things we talk about *together*?'

I had no answer. Although I imagined a future with her, I'd been happy cruising along. When we'd first met, I told her that

I wasn't interested in short-term relationships; I was looking for a marriage partner. This had almost put her off. But now I had got cold feet, and I didn't know why. Maybe it was because if I committed to something or someone, I felt compelled to see it through. If I were to commit to Anna, it meant that we would grow old together.

During that day and night, her question filtered through my mind. By the next morning, I hadn't come up with any reasons as to why we shouldn't get married — after all, I really did love her. Standing at the sink in my share house, washing the morning's dishes while she dried them, soap bubbles foaming up my arms, I asked her to marry me. She said yes.

A few months later, her extended family gathered at her grandparents' home in Sydney. It was the first time I would meet them; an opportunity for Anna to show off her fiancé. Her grandparents, post-war migrants, had made good through hard work, thrift, and family togetherness.

Anna had warned me: 'The family can be a bit overwhelming.' As we sat around in the dining area off the kitchen, I was both shell-shocked and fascinated by the noise. The conversation was like a rugby maul — each person diving in over the top of another. There were no pauses, but somehow it worked. Yet I couldn't see how to enter the maul.

After a while, Anna's grandmother touched an index finger to her temple, and, loud enough to be heard by everyone, said, 'What's wrong with him? He don't speak. Is he dumb?'

All eyes turned towards me; this was my moment. 'I like to think before I speak,' I said, straight-faced. Everyone, Anna's grandmother included, exploded with good-natured laughter, as if it was one of the funniest one-liners they'd heard. An aunty squawked, 'You gotta get in when you can, love.' I joined in the laughter. The ice had been broken.

Over the next few years, I grew to like Anna's warm, inclusive family very much, and her grandmother especially.

Sixteen years after the marriage proposal at the kitchen sink, we had achieved many of the dreams we'd talked about then: the house (although not on acreage) in a community-minded town by the coast, three healthy daughters, a thriving food garden, financial security, and a great network of friends. And our relationship had matured: over the years, the companionship and trust had increased, and we had become best friends. But romance and the sex that once went with it had died. This irked Anna greatly.

'Something's got to change,' Anna had told me recently, her tone conveying an unthinkable ultimatum.

I had sympathy for her. We'd lain in our matrimonial bed for the last two years like warmed-up corpses: a kiss goodnight, sometimes brief spooning, before rolling over onto our respective sides. Progressively, I had come to reject her advances. Sometimes I heard her cry softly in the dark. I didn't know what to do then.

So now we were going to see a relationship therapist. I had initiated the sessions; I didn't know what else to do. I was confounded by my loss of sexual interest in her. Four years earlier, I had attended a professional workshop given by a sex therapist. She had said that the most common sexual problem for long-term couples was discrepancy in desire, which existed in almost all marriages, and the couple's unpreparedness for this. I had completed a self-evaluation, which revealed my sexual desire was at least average. So what was wrong with me?

Loss of libido was a classic symptom of depression, and sure, this had been at play when I'd first started seeing Wayne. But as my depression had begun to lift, I could see that it was more than this. Reluctantly, I was forced to admit that the idea of sex had become repulsive. I was still physically attracted to women, but sex was somehow loathsome.

I was hoping that I could find some answers and we could get back to where we had once been. I had stopped the sessions with Wayne: we couldn't afford both.

Today was our first session. The therapist, a sympathetic, youthful-looking woman, asked each of us: 'Are you committed to staying in the marriage for six months while we work on this?'

We both agreed.

I was glad we had time. Anna was not suddenly going to get up and go — that was my worst fear.

Anna wanted to get straight to fixing up our sex life, but the therapist was insistent. 'You have to clear away resentments from the past if you're going to approach sex afresh.' She had us describe what it was that attracted us to each other in the early days. It was like opening a photograph album and reminiscing.

Then we got on to our grievances. Anna told the therapist that she did not feel loved, that she felt abandoned by me — I had rejected her too many times. I was plagued by resentment towards Anna, feeling that I carried the financial load and that we were not always pulling together as a team. I saw all the things about her that annoyed me. But I also knew that she couldn't be only these things. I still loved her. Yet my irritability, my stress, was filtering my view of her. I was looking through a prism of negative distortion, but how distorted was my view? I couldn't tell anymore.

IT TOOK SEVERAL sessions to wade through the hurt and resentments we both had. I didn't know how this process was for Anna, but for me it was liberating. I felt more comfortable talking about our relationship with the therapist present. She came in if one of us was accusatory; it was like we had an umpire. We were made to stay in the present and on track.

Anna said that our sex life wasn't right even in the beginning; I thought it had only been a problem since she was pregnant with our first child and we'd relocated from Canberra to Sydney. This was when I was focused on building my first private practice, and the strain of this, along with the grief I was plunged into after my mother's unexpected death, had had an impact on our sex life. I remembered when we had left Sydney, eight years before, to go on a road trip around Australia. We had camped for three months, and along the way my sexual desire had resurfaced. Our lovemaking had flowed again, even though we had a two-year-old, Ashley, with us. I'd noticed before how work and life stress could soak up my energy for sex and romance, but with these absent, things had been different.

The therapist gave us instructions on how to reconnect at home, such as catching up at the end of the day to debrief over a glass of wine, and having daily hugs. She also gave us intimacy exercises: non-sexual stroking and massage. When we tried these the next day, there were two versions of me: one that was relieved to be touching and reconnecting with this woman whose body had become a stranger to me; the other hovering close to nausea and wanting to run out of the room. It wasn't anything Anna was doing; it was me.

Some years earlier, I'd come across a newspaper article about the work of James Pennebaker, a professor of psychology at the University of Texas. It said that he'd conducted a research study in which college students were asked to write, for twenty minutes a day over four consecutive days, about a traumatic event or an emotional upheaval in their lives. In particular, they were asked to write about their thoughts and feelings on it. After the four days, the students' immune function was shown to be boosted, an effect that lasted for six weeks afterwards. Six months on, their visits to the doctor had halved. Not only had the writing exercise helped them to achieve some emotional resolution of a difficult experience, the article said, but it had also improved their physical health.

After reading Pennebaker's article, I had suggested his writing exercise to some of my clients, as a method for moving past difficult emotions. But I'd never done it myself. Perhaps I could use writing to shift my negative feelings towards Anna?

I reacquainted myself with Pennebaker's approach. Keeping secrets, he said, was bad for your health. And writing about difficult personal events helped to create meaning out of the experiences: a process of psychological reorganisation that linked cause and effect.

His instructions were simple. *Write about a past or present emotional upheaval and how you feel about it now. Write for yourself — not for anyone to read it. Don't censor what you write. Don't worry about spelling and grammar. Do it for 15–20 minutes over 3–4 days.*

So I decided to try it — or at least a version of it. Over the next few weeks, whenever I boiled with annoyance or anger towards Anna, I went away and wrote whatever came to mind. I didn't do it, as Pennebaker suggested, over several consecutive days; it was only at those times when I thought I might explode. There were often weeks between writing sessions. But each time it felt as if I was a pressure valve releasing steam. In some ways, it was better than face-to-face therapy; I could reveal any secret, write anything: no one was going to read it.

LATER THAT YEAR, an opportunity came up for Anna to travel to Europe with her mother, meeting up with her sister in London. She would also see her grandmother's homeland, which I knew was one of her life dreams. I agreed that she should go, and she set off on a six-week trip.

This interval gave me a chance to explore with the therapist my confused attitude to sex. I had a theory on this. In our next session, I told her that I'd had a lot of exposure to sexual offenders and victims in my work, and thought that there may be a connection.

She asked me more about this. I explained that I had first met paedophiles, serial rapists, and psychopaths while working as a psychologist in the New South Wales prison system in the mid-1980s, when I was in my twenties. I told her of my work with adult victims of sexual abuse and assault during my career, and my work in the last five years for the state's Children's Court, seeing children in awful circumstances. Yes, she said, this exposure could be affecting how I felt about sex.

That evening, with the memories fresh in my mind, my skin prickled with a sticky feeling and I was nauseous: the same sensations I got when I thought of lovemaking. The familiar swell of panic rose. During the night, the violent nightmares resurfaced.

And I remembered the woman who had been abducted, raped, and murdered by three men, one of whom I'd met while I was working in the prison system.

After eighteen months of working at Grafton Correctional Centre, I had been transferred to another jail in the state. It was a large, maximum-security prison sitting in the middle of a field, resembling a castle, with ramparts of razor wire. To get to my office in the innermost section, where I had direct access to the inmates, I needed to walk through eleven doors and gates, and all of these, except for my office door, had to be unlocked by a prison officer. It took fifteen minutes to walk a distance of 200 metres.

The arrest of the suspects in this particular case had led to intense media attention. In my first month on the job, I found myself sitting at a small table opposite a frightened young man, not all that much younger than me. I had been told he was 'not coping', that he and his family had been getting death threats, and that I was to help him in whatever way I could.

We were seated in the protection unit, sitting on characterless plastic chairs. Inmates sent to this unit were there for protection from the wider prison population because of the nature of their crimes, which made them a target. Child molesters, known as 'rock spiders', were usually placed 'on protection'. Although this man was awaiting trial, there was no doubt among the prison officers I spoke to that he was guilty.

Grafton, a country jail, had been like a retirement home for many of the 'lifers': inmates who would see out their days in incarceration. I'd read through their files — as horrendous as some of their acts of violence were, the crimes had been committed well in the past; their intensity was muted by time. But now, I was sitting across the table from a man who had, I believed, recently committed a horrific murder. His mop of dark hair, slight body, prison-green shorts, and tan-coloured shirt gave him the appearance of a toughened schoolboy rather than a killer. But if the reports were correct, he was ruthless and vicious.

The room we were in had no window, flooded instead by fluorescent light that made everything look washed out. A door banged off to my right, where the nearest prison officer was stationed. The air tasted of confinement, and every so often I caught the unsettling whiff of sweat and testosterone. I saw the man's two accomplices through the glass wall behind him. One was prowling up and down in the small, sparse dining area. He stopped, turned, and sent a cold stare in my direction. I wasn't sure if he was trying to intimidate me or if that was how he always acted. The other was scribbling diligently in an exercise book at the table, as though doing his homework. I was told later that he was making notes for his trial.

The young man sitting in front of me was sniffling. As he spoke, I listened quietly and eased back in my seat, trying to present a picture of professional calm.

In the almost two years that I had been working in the prison system, I had been struck by how sad most of the inmates' lives had been. Repeatedly, I had asked myself, *If I had been brought up in the same circumstances as these men, would I have made some similar choices?* They elicited my empathy but also a feeling of hopelessness: how much would my puny input help when they were so damaged? And yet, I tried to be one caring human being in the system.

But in this case, I was having difficulty. I could never imagine making the same choice as the man sitting in front of me. Swelling within me was a torrent of disgust. I worried that I might vomit. I reminded myself of my role here: not to judge; to see the person before me, and help if I could. I pushed down the welling repulsion. I willed myself to lean forward, to bring my face closer to his, and even to feel sympathy for him. Sweat glistened on his upper lip.

After this meeting, we had one more. For confidentiality reasons I cannot say what we spoke about in either of them, but he seemed helped by having someone neutral to talk to. After that, he no longer requested my presence; he was swept away in daily court appearances and meetings with his lawyer. I was relieved.

After a year at the jail, the small world of the prison had expanded in my mind, while the world outside had become small. When I would meet new people 'on the outside', I'd ask myself, *What do they want from me? Can I trust them?* Mentally, I'd become an inmate.

I resigned and moved to my old hometown, where I commenced my PhD, and eventually I took a position in Canberra, where I met Anna. I seemed to forget the prison experiences, and things were good for a while. But after Anna and I relocated to Sydney, I worked for a consultancy that offered psychological debriefing in workplaces after critical incidents such as robberies, assaults, and industrial accidents. I was thrust into situations that reminded me again of how much harm was caused by criminals, and in particular, it seemed to me, by men. One day, I said to one of the consultancy partners, 'You know, Helen, most of the abuse and violence in the world is caused by men. I'm ashamed to be male.'

There was silence on the other end of the line: was she taken aback by what I'd said, or was she in agreement? Then she replied, 'We can only change what we can,' and gave me the details of the new call-out she had for me.

Now I wondered: was I carrying within me a web of memories of male aggression? Had it become welded, for me, onto the sex act? Was sex a form of aggression, a violation? Was this what I carried into lovemaking?

But my criminal work wasn't the only thing I connected with my distaste for sex. I thought back to my role in carrying out family reports for the Children's Court. Children were taken from their parents when there was abuse or neglect, or when the parents were not coping because of substance dependence, mental illness, or intellectual delay. I often assessed children of similar ages to mine.

I remembered the woman who'd had seven children to two partners. All had been removed from her care, and it was the eighth child — a baby to a third partner — whose future the court was considering when I was asked to submit an assessment. I had stated in my report that I doubted she was capable of looking after her child adequately. It wasn't long after the court's determination that the baby be permanently removed from her care that she and her partner came up to me in a local cafe. Given the circumstances, I was surprised when they beamed, greeting me like a friend. 'Janice is pregnant,' her partner said. 'We're so happy.'

I thought, *Another child that will be pushed into the world with the hope, at best, of being parented by a caring stranger. Look at what sex does.*

MAKING THIS LINK between my professional exposure and my attitude to sex gave me greater perspective. I could see now that the source of the disgust was old memories, body memories. I could gain release from them, some separation. A healing could begin. I'd made an important discovery.

When Anna returned, we seemed to be nicer towards each other. I told her about my discovery in regard to sex. I had something I could work with. She digested the news quietly.

Slowly, over the next few weeks, my sexual desire started to come back. Time apart from Anna, having our resentments out in the open, and making sense of my relationship to sex had all helped. Something I'd thought had been lost had been found. Over the coming weeks I made tentative advances — a hug, a kiss — not to be sexual, but to express a newfound affection.

One day I came across Anna standing by the sink, peeling potatoes. The late-afternoon sun was squeezing through the blinds, illuminating the stainless-steel benchtop. From the pile to her right, she picked up a potato covered in dirt, rinsed it, and began to peel it with a paring knife. The peeled potato was dropped into a pot of water. These actions were repeated. She was quick: her arms rotated back and forth.

I came up behind her. My groin folded around her warm bottom. My chest pressed against the points of her busy shoulder blades. I rested my hands on the delicious curve of her waist.

Yet I was met with a stiff back. Her arms didn't stop. I nuzzled my face into her musk-scented neck, but her head tilted away. She didn't say anything, her shoulder blades still pumping like pistons. Uncertain, I hovered longer; maybe my loving presence would soften her.

'David, I need to get these on the stove,' she said, turning to face me, her hands gripping the U-shaped handles of the pot.

'Why don't you want me around you?' I asked.

'I can't compete with the pictures in your head,' she said, referring to the images I had told her plagued me when thinking of sex.

'But those are gone now,' I replied impulsively. They hadn't gone, but they were loosening their grip; I had hope. What else could I offer her? I was throwing out a lifeline to stop her from drifting away.

'It's too late, Dave. I don't have any desire for you anymore. I've switched off. Too much has happened.'

I had no reply. I walked away. I was confused by the lack of softness in her face, her steely logic, the sense of a cause lost. Now I knew what the thud of rejection felt like, what Anna must've felt.

6

SOON AFTER ANNA'S rebuff, I woke one morning with the thought, *Is there a psychological insurance policy for life shocks — a policy that could restore me to how I was before the shock?* What if the things largely outside my control, the ones that had knocked me around — the post-traumatic stress, Anna's feelings towards me, Dad's death — could wash over me and I could bounce back, instead of being left reeling?

In cognitive behaviour therapy (CBT), unhelpful thoughts are challenged and changed to realistic, more useful ones: a process called cognitive reappraisal. Wayne had challenged my view of myself as a failure, encouraging me to see that I could embark upon a new career phase that drew upon my past experience. He had also helped me identify behaviours undermining my recovery. His instruction for me to stop watching television news had been aimed at reducing the amount of human suffering and vicarious trauma I was exposed to: a way of lessening my fight-or-flight response and, consequently, improving my sleeping patterns.

In my PhD research years before, I had developed an extension of the CBT approach to assist musicians in managing their performance anxiety — stage fright — so that they could replicate what they'd done in rehearsal in front of an audience. I'd consulted sports psychologists at the National Institute of Sport in Canberra about performance under pressure, read the latest research on mental rehearsal, and developed imagery techniques for performing artists. It had worked: most performers in my study felt calmer in performance, their heart rates were lower than prior to training, and they made fewer mistakes. I'd even written a self-help book, which ended up being published in four countries, describing the strategies I'd developed. It had been used in university courses, and I'd been asked to deliver many workshops to performance groups. And I'd drawn on the same strategies later on, when asked to help medical doctors taking their oral exams, nervous truck drivers taking their driving tests, and anxious public speakers. I used the approach myself when speaking in front of large audiences, and in radio or television interviews. So the idea of changing the mind with a training technique, such as mental rehearsal, wasn't new to me.

But despite the sessions with Wayne, I could still be overcome by tsunami-sized emotions. The cognitive defences I propped up would be washed away like fences in a flood. I needed a new approach.

When I was nineteen, I had taken a course on transcendental meditation (TM) with my mother, and for the next three years I meditated twice daily for twenty minutes, and did some of the more advanced TM training. But as the years went by and I got busier, my regular practice lapsed. Now my thoughts turned towards meditation again: could I change my mind in some way — maintain an inner calm as I had learnt to do before with TM, but in the face of my challenging circumstances?

And not only meditation: mindfulness was a concept being mentioned more and more often in the mainstream psychology literature I read and at the professional gatherings I attended. I didn't fully understand it, but I did know that it was a different approach from CBT.

I had read a professional book on the subject, *Mindfulness-Based Cognitive Therapy for Depression* by Zindel Segal, Mark Williams, and John Teasdale, the year before I'd closed my practice. They had devised a program to stop depression sufferers from experiencing repeated episodes: relapse prevention. They drew on research which suggested that while cognitive therapy was great for helping people get out of a single depressive episode — especially in conjunction with medication — it was less successful for people with long-term recurrences of depression, especially when these were due to difficult, unchanging personal circumstances.

Their eight-week group program drew on the work of Doctor Jon Kabat-Zinn and the mindfulness-based stress-reduction program he had developed at the University of Massachusetts Medical School years before. I'd come across Kabat-Zinn's approach for sufferers of chronic medical conditions briefly during my postgraduate training, when we'd had a seminar on the management of chronic pain. Kabat-Zinn offered a way for patients to live more contentedly. The aim was not to change their condition, but to change their relationship to it: to change their psychological response to physical pain.

Segal and his colleagues found that their program reduced the likelihood of those who'd had three or more previous depressive episodes falling back into depression by 50 per cent, when compared with the conventional relapse-prevention methods of medication and CBT. This was a startling result.

I pulled *Mindfulness-Based Cognitive Therapy for Depression* off my bookshelf and re-read it, to find out what was different about their approach. They viewed conventional CBT as a 'doing' approach; it tried to problem-solve a way out of depression, anxiety, or other undesirable states of mind. But if the circumstances triggering the depression or anxiety were unchanging and outside the sufferer's control, how could one problem-solve their way out of that? Mindfulness, on the other hand, was a 'being' approach, they said; it encouraged acceptance, a sense of equanimity in the midst of a person's circumstances. This took the pressure off having to change disturbing thoughts, uncomfortable feelings, or unpleasant bodily sensations such as pain. The mindfulness tradition came out of Buddhism.

I had used their approach with one client who had a history of depression, not long before I'd closed my practice. He was concerned that he might have a relapse. I told him that it was a new approach — he'd tried everything else — and he agreed to give it a go. We started with the raisin exercise: you take a raisin in your hand and look at it as if it is the first time you've ever seen such a thing. You rub it between your fingers to get the texture and weight of it. You smell it. Then, you put it in your mouth and eat it slowly, taking in every sensation about it that you can. I gave him the instructions for the breathing meditation described in the book. He did the exercises and the homework for four weeks, but then he had to discontinue his sessions due to personal circumstances. He liked the exercises, but unfortunately it was too early to observe how much of a difference they made.

I had also attended a daylong workshop given by a clinical psychologist who spoke of mindfulness in the context of treating anxiety and depression. He had found it useful personally, saying that he'd always had a very active mind and needed something to slow it down. He had invited a PhD student, and she spoke of her research with adolescents who had obsessive–compulsive disorder. She'd trained her subjects to see their obsessive thoughts as the flotsam and jetsam of the mind — things they didn't need to place importance on or act upon. She found that her adolescents reduced the ritualistic behaviours they usually engaged in to relieve their anxiety.

The clinical psychologist gave us each a CD with his recording of a breathing meditation, a walking meditation, and a body scan. The walking meditation encouraged the focus of attention to the feet's contact with the ground, while the body scan cast the mind's eye to the sensations in every part of the body. I tried these out at home, but his brief teaching of the mindfulness concept and the exercises didn't do much for me. Still, I remained excited about its possibilities, and had always meant to look into it further.

Could mindfulness and meditation — whatever the connection between them was — be the insurance policy I was looking for? If I could get the hang of them, they seemed to promise mental composure in the face of my circumstances.

THE CEILING DREW my gaze up to a diamond-shaped skylight at its apex. I was in a vast circular building at Sydney's Olympic Park. Before me was a broad stage, behind which hung long paintings of Buddha-like figures in yellow and gold. An elevated seat dominated the centre of the stage. To the right and left of it, monks and nuns in maroon, grey, black, or brown robes sat cross-legged on cushions.

I was sitting in the middle of a row close to the stage. I didn't know anyone. The people around me looked decidedly ordinary. Some had a glassy-eyed look of devotion; many talked excitedly, filling the auditorium with a human hum.

I had come for a five-day course, Stages of Meditation. I'd seen it advertised a few months before, and it sounded like it would take the attendee progressively through the different levels of meditation practice. And it was being taught by the Dalai Lama: a man I admired for his compassion and wisdom.

There was movement behind the monks and nuns. A slightly stooped figure, wearing square-rimmed glasses and maroon robes, emerged from behind them. It was the Dalai Lama. He bowed to monks and nuns on his way to the front of the stage, taking their hands in his, sometimes touching his forehead to theirs. Great respect was evident among them. When he got to the front, he acknowledged us with palms raised together. There was loud applause. He climbed onto the seat and sat cross-legged, securing a microphone, the kind that pop stars and motivational speakers wear, to his head. 'Good morning, everybody.'

We had begun.

After some introductory remarks, he said, 'We have this marvellous brain, with that special intelligence ... Intelligence and education do not necessarily bring inner peace. Throughout human history, people have been trying to find different ways of obtaining inner peace, particularly in difficult situations ... Hope is necessary to face helplessness or difficulties beyond our control.'

Yes, I had come for hope.

By day's end, however, it had become clear to me that this was not a course on meditation but an in-depth teaching on an ancient Tibetan Buddhist text, a translation of which we had in our booklets. I found it difficult to remain alert during the Dalai Lama's long monologues in Tibetan, followed by the translator's soporific voice. But it had been a huge effort to arrange seven days away, including accommodation and flights. I wasn't going to leave.

I was staying in cheap dormitory-style accommodation in an ex–army barracks within walking distance from the venue. Everyone at the barracks was attending the teaching. This gave the place, despite the spartan surroundings, the feeling of a retreat. My room housed twenty men. In the bunk adjacent to mine was a retired farmer with a bushman's beard and a disarming smile. He identified as Buddhist. 'My missus is not into it; she thinks it's mumbo jumbo,' he told me. 'But she doesn't mind. "As long as you do your business in the shed," she says. The shed's pretty well set up.'

We laughed. I loved the vision of him retiring to his shed to chant and meditate, and I guessed that meditation wasn't the only reason he liked to spend time there. I told him that the course was not turning out the way I'd thought; I'd come to learn mindfulness meditation. He explained that what I was after was called calm-abiding meditation.

On the second day, I let go of trying to understand everything the Dalai Lama said; instead, I closed my eyes and allowed the sound of his voice to resonate through me. It was surprisingly soothing. At the end of the day he answered audience questions in English, and I perked up. He was funny, injecting humour at surprising moments. I was touched by his compassionate nature, and impressed when he spoke of his cooperation with neuroscientists who were investigating contemplative states.

The next morning before the teaching, I went in search of good coffee, entering a nearby hotel. As I sat sipping it in the foyer, I noticed that a group of people who looked to be from my course were forming a semi-circle off to one side of the reception desk. A woman or a man — I couldn't quite tell which — in maroon and yellow, with a shaved head, stood at the head of the line. Next to her was a man in a wheelchair. They seemed to be waiting for something. *This looks interesting*, I thought, so I joined them.

Soon, men in dark suits and earpieces appeared from a doorway behind reception. Then we saw the Dalai Lama materialise, following them. When he approached the group, he came up to each person, greeting them in turn. Soon he was up to me. He took my palms in his, and, looking into my eyes, said something I didn't catch. Then he moved on.

I couldn't believe it — I'd just missed what he'd said to me! Of course I was pleased with this chance meeting — how many of the other attendees would have liked it? — but I was kicking myself: had I missed something insightful, something of personal significance, or had he given me a kind of blessing that he gave to everyone? Afterwards I rolled the sounds he'd made around in my mind, hoping they'd gel into something meaningful, but they didn't. I hoped my chance to gain some understanding of the mechanics of inner peace wasn't passing me by.

EACH MORNING, MONKS and nuns from different Buddhist traditions gave guided meditations in the auditorium. A nun from the Thai Forest Tradition instructed us to attend to the sounds around us without holding on to any particular sound and without making judgements. Another teacher asked us to count our exhalations up to ten and then repeat this cycle. If we lost our way before we got to ten, we were to start back at one. A third teacher asked us to visualise different colours on the inhale and the exhale. There was no one way to meditate, it seemed, and no one had yet spoken specifically of mindfulness. I understood that all of these techniques were connected to enhancing mindfulness, but I wasn't sure how.

During the long lunch breaks, we could line up for an interview with one of the nuns or monks. There was always a queue, but on the second-last day I joined it. I wanted to ask what I could do as a regular meditation practice at home.

Finally, I got to sit across from a middle-aged monk who said he was a youth worker. In answer to my question, he told me to focus on the image of my guru. I didn't have one, I said.

He suggested Chenrezig, the embodiment of compassion: an image associated with the Dalai Lama. I should begin with three prostrations: he demonstrated by joining his hands together in prayer pose, throwing himself onto the floor, and repeating this twice more, panting heavily. This was alarming; I couldn't see myself doing this at home. I looked around, feeling a little embarrassed, but no one seemed to take any notice. After doing that, he said, I was to visualise Chenrezig, and with time the image would become clearer in my mind.

So, yet another meditation technique. I felt a surge of irritation. This was so confusing. I wished there was a course that a beginner like me could do that covered all the basics. These Buddhists needed to get their act together — to agree, somehow, on what the essentials were. I sensed that the Buddhist teachers had a depth of experience from which I could learn, but I missed the clarity that the clinical psychologist had exuded in the workshop on mindfulness and meditation.

Although not quite convinced by the monk's instructions, I went to one of the stalls selling icons and bought the smallest and cheapest Chenrezig statue I could find. I would give it a go.

That night, after eating out, I came back to find most of the men hunched around the small television in our tiny living area: it was one of the big footy games of the year. They were barracking and cursing. I smiled — and here I was, wondering that I might have been getting myself into some kind of Buddhist cult, with prescribed forms of behaviour. I joined them for a while before heading into my dorm.

A young builder from New Zealand occupied the bed above me. He got up at six o'clock each morning to do prostrations on the porch; in my half-sleep, I would hear the thumps on the floorboards. A general practitioner, with three children similar in age to mine, slept in the bunk above the farmer. The next morning, he talked of the need for health professionals to use more 'heart communication'. He taught meditation in his local community and to some of his patients.

I liked these men; they were straightforward and seemed genuine. I was the novice, but I didn't feel excluded. They didn't show any trace of trying to force their beliefs onto me.

That afternoon, as the rows of seats were being vacated at the end of the course, I turned and saw an older man with a stringy frame and silver hair. He was looking vacantly but calmly at the stage. I caught his eye. 'Hello. How's the week been for you?'

'I've got a few things I can take home,' he said, nodding.

'Where are you from?' I asked.

He named a town out west — an area that I knew to be affected by drought. He said he was a farmer.

'Things have been hard lately,' he told me. I saw pain flash across his face. 'I've got to handle stress better. I'm glad I came.'

I smiled. I wasn't the only one looking for an insurance policy.

UPON MY RETURN home, I felt renewed, despite not yet having a definite idea of how to undertake mindfulness meditation. With a new commitment, I took the Chenrezig statue, a black metal Buddha we'd bought on holiday in Thailand, an incense burner, a blanket, and some cushions, and set myself up in the girls' cubby house.

The cubby had been an afterthought to the house. Once we moved in, Anna said we needed one. Ashley and I had sat down with a book of cubby-house designs and agreed on one we liked. It was classic, a colonial farmhouse with a gabled roof and a long verandah. I imagined the girls sitting on the verandah in their yellow and pink chairs, looking out over the garden and having tea parties with their friends. Anna was happy with the design but, being practical and farsighted, wanted it to be large enough for a double bed so that guests could stay, while still leaving room for the kids to install their toy kitchen and to play school. And so the cubby was expanded into a pint-sized garden studio tall enough for an adult to stand in. We'd painted it the same colours as the main house and attached the same grey tin roof.

I made the cubby house my place of refuge, away from the commotion and demands of the family. I tried different ways of meditating. I visualised the Chenrezig image as best I could, peeking at the statue every now and then to remind myself of its particulars. I tried attending to the surrounding sounds, not placing emphasis on any individual one. I tried counting my exhalations. I tried visualising a dark colour while breathing out and white light while breathing in.

Gradually, over several weeks, I increased the length of time I was able to sit in meditation. And gradually, something changed in me. I began to feel motivated. I weeded the garden and stopped reading the newspaper from cover to cover.

The next time I met with Wayne, I reported my renewed vigour. I was sleeping better and exercising again. And I had made a friend.

I had met Nick, who was around the same age as me, at a barbecue. He'd played classical guitar in his youth, like me, and wanted to do so again. He was eager, with a grin that reminded me of a cocker spaniel. We agreed to meet up and play some duets. The day after we met, I pulled out my old sheet music, warmed up my fingers with scales and arpeggios, and went through the short pieces I used to play.

Before long, Nick and I were playing together regularly. Nick was a hustler, prodding me into action with his enthusiasm. Soon we started a series of fundraising concerts, held at a local community hall, inviting other musicians to perform. The duets with him and the contact with other amateur musicians brought me pleasure.

I began to have the energy for other new projects too. For years, I had been mulling over an idea for a book on risk-taking: why people did or didn't take risks, such as embarking on a new relationship, or ditching an unfulfilling job. Now I picked up books on the subject, started making notes, and worked out a rough chapter outline. I came across a book, *The Artist's Way* by Julia Cameron, that recommended stream-of-consciousness writing: 'morning pages'. So I added this exercise to the end of my meditation session, as a way of generating ideas for the risk-taking book. I hadn't written about Anna for a while, but my thoughts and feelings about whatever was happening with us also began to slip into the jottings in my journal.

It seemed that life had finally turned around, and I dared to hope that my 'insurance policy' was working. To reward myself, I decided to go on a seven-day group trek along a desert mountain range. I hadn't walked in wilderness with a backpack since my twenties. During that week, I exhilarated in the sense of freedom: the momentum of walking without children tagging along, and the splendour of camping under big desert skies without a building in sight. Life had taken on a new sheen at last.

I returned to civilisation in September, to hear news of the Lehman Brothers collapse. Global stock markets were plummeting. It was being referred to as the global financial crisis. At the same time, I noticed how low our bank balance had become.

7

OVER THE LAST year, ever since Dad's death, I'd hankered for someone to look up to — a parent to listen to my miseries, give me an accepting hug, and tell me something wise. Dad wouldn't have been able to do anything practical for me, but he could've provided these things. After his death, I realised what a vacuum there was; as the eldest of my siblings, I was now at the end of the family line.

Being the executor of Dad's will, I'd had to deal with his accountant, his solicitor, and my extended family. The Department of Veterans' Affairs had also contacted me: they'd made Dad a payment before becoming aware of his death, and it needed to be refunded. I had to make sense of his pharmaceutical and nursing-home accounts. A final tax return was required; Dad continued to pay tax long after he was gone. I debated dollars and cents with clerks on the phone as if Dad were still alive. The people I dealt with were only doing their jobs, but each conversation, each occasion I went through his documents, re-ignited my sense of loss, and the ache in my chest would restart. Every time, it took several days for the veil of grief to part.

As the first anniversary of his death approached, I got an anxious feeling, as if something awful was about to happen. I called my sister and spoke to her about it. She said she didn't feel apprehensive and that it was okay. She was right, of course. Following the anniversary, the feeling eased, and I could finally stop thinking about him so intensely.

When I had first stopped work, it meant that I could help out more with the children, relieve the domestic pressure on Anna, restore my health, visit Dad when he was ill, and look after his financial affairs. It also meant I had time to extend our property portfolio.

I enjoyed dealing with numbers, and with property managers, builders, and tradesmen. With men in work shorts and boots, there was no need to step around sensitive feelings. And I liked talking about things I could see and touch, things that would last — a contrast to psychology, in which most of the people whose lives I'd helped I would never see again.

At Dad's funeral service, a cousin had presented me with an extensive family tree he had drawn; there was a line of builders going way back. Our earliest Australian male ancestor, originally a convict, had made good, building churches and hotels, and eventually becoming the mayor of Waverley, in Sydney. He even had a street named after him in Bondi. Dad had been handy at making things too.

I'd always enjoyed looking at buildings, identifying their architectural periods and features. It was a passion that Anna and I shared. I'd renovated my first house on my own, before my marriage. Dad had shown me how to do the electrics and lent me his prized tools. So my cousin's revelation of our architectural line had a neat symmetry about it.

In 1993, one of Anna's relatives had told me how he'd built his wealth through property investment, and how we could do this too. 'You do your psychology thing, which is what you really want to do, and let your assets build up on the side. You'll hardly have to think about them,' he had said.

Anna liked the idea of investing too. With her relative's guidance, we nervously made our first property investment, and then bought our home. By 2002, with three young children, and conscious of the uncertainties of self-employment, I wanted to expand our portfolio. By now, I was more enthusiastic about devoting time to this than Anna; she was developing her business interests, post-babies.

By 2007, with the escalating property market, we had become, on paper, wealthy. During the unfolding of this ridiculously easy path to success, I had happily mentioned to friends and family — trying not to be evangelical — that they could do it too. A few took up the challenge; most did not, and I felt sorry for them: tied to nine-to-five jobs with no other options.

As our wealth grew, we seemed to want, or even need, things that we had never needed before. I was keen on building a beach home. I wanted a retirement apartment by the sea for Anna's mother and stepfather; overseas trips for us and the children, to have new cultural experiences; and the children's future education expenses (no matter what or where they wished to study) to be met. I imagined stopping work altogether one day in the not-too-distant future, devoting my time to music, writing, or 'good causes'.

We had a multi-million-dollar property portfolio with lots of equity, but it was not self-supporting. The rental returns did not cover the outgoings — the largest being interest repayments. We had aimed for capital growth with the idea of selling down at some point to make it self-supporting and put cash in the bank. But after 2006, when I could no longer work, my income stopped. At first we were fine, relying on savings and drawing on the equity in our properties. Yet the reserve bank raised interest rates throughout 2007, and lenders raised their rates faster than the reserve bank, bleeding our funds for loan repayments. It kept going until early 2008, bringing the property market to a standstill. As fear gripped the nation with the onset of the GFC, and the market slid in value, people stopped buying in the face of the unknown, and credit dried up.

With the economic recession now a reality, we had little other income. Forces outside of our control had us in a bind: we couldn't sell, we couldn't refinance, and we couldn't meet our repayments. A prominent economist interviewed on television said that property values would fall by 40 per cent before the crisis was over. Such a fall would wipe out any equity we had left and leave us with loans we couldn't repay. We could not see a way out.

I decided that I would have to return to work. After all, I had gotten much better. But the more I contemplated a return, the more I got that old sense of dread.

By December 2008, our funds had gone. We asked family members for loans so that we could put food on the table while we waited for our first property sale — in a dead market — to happen.

The benefits of the therapy, swimming, music, the teaching with the Dalai Lama, and meditation seemed to vapourise. My jumpiness and nightmares returned. I was close to shutting down: so worn out by dealing with the calamity that was unfolding and with my deteriorating mental health. I wanted to walk away from it all, even though I knew that this would help no one.

One afternoon, I was driving back from a new construction — started before the GFC had taken effect — after meeting with the architect and the builder. We'd argued over one of the contractors, who I felt had overcharged. Out of nowhere, I was hit by the strongest desire to drive off the highway at one hundred kilometres an hour, down a steep embankment — hopefully to a quick death. It was a vicious, uncaring world, and I couldn't meet it head-on anymore. The past me, the strong me, the one that blinked after a setback and then got on with things, was gone.

'Stop! Stop!' I shouted, to that part of me that was now being very scary. I needed it to stop carrying on like this, with this craziness.

But the steering wheel coaxed me to the left, off the side of the road, the deep embankment promising deliverance.

I tried reassurance: *You'll get through this. This feeling will pass.*

I tried alarm: *Imagine how distraught Anna and the kids will be*. I saw their shocked, unbelieving faces upon learning of my death. I saw my friends' horror.

I tried compassion: *I would be letting everyone down; it would cause immense heartache.*

I tried a warped logic: *What if it doesn't work and you end up crippled, and can't make a second attempt?*

I tried the radio: I had to get my brain out of this groove.

With the radio jangling my thoughts, I managed to hold it together until I arrived home, trembling, and collapsed onto the bed. I was afraid of myself — what was my brain doing? When would the next suicidal urge swoop down upon me?

I knew from my work that when people are overwhelmed by sustained physical or emotional pain, they can lose hope that the pain will ever go away, seeing no point in sharing their thoughts with others. *The world would be better off without me*, they think.

I was sliding down this path.

I summoned my courage and told Anna about the experience on the road, explaining how worn down I had become and how I couldn't deal with the financial stresses anymore. She was taken aback, almost speechless. 'I'm sorry you're feeling this way,' she said.

I caught up with Lily. She reminded me of all the people who cared about me. She asked me to contact her at any time, even if it was only to go for a walk when I was grumpy.

I wanted to have a proper talk with Ian: not one of our standing-at-the-car updates after a swimming-squad session. I wanted to know, specifically, if he thought I should take antidepressants. I had taken St John's wort for short periods over the last two years when feeling low, and this had helped.

St John's wort was a herbal medication that many trials had shown was effective in the treatment of mild to moderate depression; in fact, it was as effective as the commonly prescribed selective serotonin re-uptake inhibitors (SSRIs) in these cases. It didn't have the side effects of the SSRIs, although it could have a few of its own — thankfully, none of which I had experienced.

It took two weeks before Ian and I could catch up for lunch; I hadn't told him what I wanted to talk about. He'd been preoccupied with family and work, he said when he arrived at the cafe. 'I need to give friends greater priority,' he said with a rueful smile.

We sat at an outside table, where we could smell the nearby ocean. The sandaled foot of his leg, which was crossed over the other, poked around in the air like the snout of a sniffing dog.

I described my experience the other day, driving home. I said those thoughts had hung around for a few days afterwards; now they were gone, but they didn't seem far away. He asked how readily I was able to get out of a low mood. I said that a swim or a music session was usually enough to do it. St John's wort made a difference.

'It sounds more like you're reacting to circumstances around you,' he said. 'If the St John's wort is helping, keep this up, but be consistent with it.'

After opening up to Anna, Lily, and Ian, my suicidal thinking seemed ridiculous, embarrassing; I might have been exaggerating it, I tried to tell myself. But underneath I knew that this wasn't the case, and I was still apprehensive — not at all confident that the urges wouldn't come back. Anna checked in with me every now and then, asking how I was faring.

The suicidal episode was constructive in one way: if the thought of going back to work was bringing on such extreme notions, I shouldn't be contemplating a return at all.

My other option was to make a claim on my income-protection policy, the one I'd been paying into for all these years. I'd never thought the policy was necessary (if we did have to draw on it, it would only be because of an unexpected physical injury or illness), but our financial planner had encouraged it strongly. I thanked him for it now.

I had considered this option over the previous months as our money ran low, but I'd wanted to cope using our own resources. And I hadn't wanted to put myself through the insurance wringer. Many of my past clients had been made worse by the sometimes brutal insurance-claim process and in dealing with young, naive insurance case managers. But recent events had forced my hand.

Three weeks later, I visited my regular GP, Doctor Sunbury, and told him that I was not coping. This felt awkward; I'd only talked to him about my physical ailments before. I showed him the results of a psychological questionnaire I'd completed, which measured compassion satisfaction, burnout, and secondary trauma. I was in the average range on the first two scales, but high on the secondary-trauma scale. He helped me to complete the insurance claim, and I sent it off.

The insurer organised an appointment for me to see a psychiatrist, over an hour's drive away. The company was also good enough to start payments immediately, while I went through the claims process, so our family could survive for the time being.

On the day of my appointment with the psychiatrist, I walked into a corporate-style waiting room, in an office suite where specialists rent rooms for the day. I was on time, and sat down to wait. I saw a middle-aged man with dyed-blonde hair walk out of a room. He strolled straight past me without a glance and disappeared into the lift. Half an hour later, he returned holding a cup of coffee, and some time after this, he called me in.

He made no comment about his lateness.

Doctor Waverly sat down on the other side of a large desk. There were two seating options for me: a low chair and a high chair. The high one, he said, was for elderly people who had difficulty getting out of chairs. I opted for the low chair, which meant I was looking up at him.

He started up a friendly chat about professional matters, and about an upcoming annual mental-health conference — would I be attending? After ten minutes, as if only just realising why I was there, he shifted into a professional tone and said, 'We'd better get on with it.'

His questions were extensive, and he recorded my answers in his handwriting, on a form with spaced headings. I was nervous: my family's financial wellbeing was riding on the outcome of this, whereas for Doctor Waverly, it was just another report. I emphasised my exposure to vicarious trauma, seeing this as the primary cause of my deterioration.

Yet after the interview was over, I thought of things I *should* have said. But it had been impossible to keep everything in mind with his questions rattling me along, spurring on my nervousness. Afterwards, however, more memories of direct exposure to trauma surfaced, as though Doctor Waverly's questions had stirred a mental pot: the heavier ingredients swirled to the top.

I remembered the Children's Court assessment I had conducted in a woman's home. She'd had her two young children removed from her recently. I was to provide a report detailing her side of events, giving an opinion on her psychological health and outlining what I thought needed to happen if she were to have responsibility for the children again. She was a single parent. I had settled down in her lounge room, where she had invited me to sit, seeming ready to talk things through, when she grabbed a kitchen knife and made to lunge at me. Then, apparently thinking better of it, she put it down. Instead she began to rage at me, vitriol pouring from her mouth.

I stood up, extending my hand in the stop position — like cops do — and said, 'Stop. Stop.' Trying to cut through her stream of venom, I said that I would terminate the interview if she didn't allow me to speak. But it was like trying to put out a fire with a flammable liquid; her face reddened and contorted even more. So I backed out the front door to the sound of her screams, which travelled all the way down her garden path. She continued ranting, following me and walking out onto the road, as I drove off.

Now I felt shocked that I could have thought at the time that this was an incident I should take in my stride, just another event in a normal workday.

If only I could re-do the interview with Doctor Waverly. But the insurance-claims system was a one-shot game.

WHILE WAITING FOR my claim to be determined, I attended a course given by a Tibetan Buddhist instructor that a research scientist at the Dalai Lama's teaching had recommended to me. It was inexpensive and practical. He taught calm-abiding meditation.

My sessions in the cubby house became more defined. Now I started with a brief body scan to get in touch with my inner sensations. I imagined that I was breathing in white light, conceiving of it as cleansing and uplifting. In the pause before exhaling (the 'calm abiding'), I pictured red light nourishing and dispersing throughout my body. On the exhalation, I visualised dark-blue light releasing all negativities. All the while, I kept my eyes partly open, with my gaze turned downwards. At last, I thought, I had a solid mindfulness-meditation practice.

THE INSURER RECEIVED Doctor Waverly's report a few months after the interview, and sent me a copy. In it, he agreed that I was suffering from anxiety and depression — although I did not qualify for the diagnosis of post-traumatic stress disorder, he said, because I had not been directly threatened. He concluded that I was not fit to work in my usual occupation.

The insurer accepted my claim, and a pattern of completing monthly claim forms with my GP commenced. The payments covered our living expenses, but little else.

I faced each weekday morning with fear and loathing, going into the home office like a miner entering a pit while knowing that the dust was damaging his lungs, even shortening his life. I dealt with the lawyers' telephone calls and the creditors' threatening letters, and juggled our dwindling funds between bank accounts to pay the latest bill, credit-card payment, or loan repayment — if we could.

The unrelenting nature of these dealings caused a constant chatter in my head, leaving little space for positive thoughts. I wanted to run away, even desert the family, to 'go bush'. I imagined that this would stop the noise, but I knew in reality that the relief would be temporary.

One particular day, a Tuesday, had been a series of financial setbacks, and I needed some quiet time. As a rule of thumb, we liked to get the children ready for bed at eight o'clock on school nights, with lights out by eight-thirty. I usually read to Amelia, while Emma was keen to sleep, and Ashley was allowed to have the lights on a little longer and do her own thing, quietly. If we didn't follow this routine, the mornings became dominated by goading grumpy children to get ready for school.

I'd gone out for a long walk that evening to try and clear my head. When I came home, I wandered downstairs and found Anna, Emma, and Amelia cuddled up on the couch, watching television. It was almost eight-thirty. 'Why aren't you in bed?' I snapped.

'Can we just watch to the end of the show?' Emma asked.

I felt my blood rising. What was Anna doing? 'Turn the bloody TV off,' I barked.

The girls pressed in closer to Anna.

'David, you're overreacting,' Anna said.

'Get into fuckin' bed now,' I shouted, glaring at the girls.

Their faces went white and their eyes widened, but they didn't move.

I picked up the remote, flicked the television off, and threw it on the ground. It made a thwack as it hit the hard surface and came apart, splaying across the floor. I was saturated with rage. I turned to the large, old-style television, sitting at waist height on its wooden stand, grabbing its sides and yanking it forwards and backwards. 'Why don't you do what you're told!' I yelled, still facing the television. My right foot lashed out, kicking the stand. A pain zinged up my leg.

I turned and saw Anna and the girls scampering up the stairs. At last!

The pain in my foot was excruciating. I went to sit on the verandah to cool my temper and my foot. When I came inside, the girls were in bed with the lights out; Anna was talking softly with Emma. I left them alone.

The foot pain continued throughout the night.

It was a subdued atmosphere the next morning as the kids got ready for school without fuss. I felt sheepish.

During the day, I saw Doctor Sunbury and told him what had happened. He examined my foot and said I'd broken my second toe. He bound it to the next toe to act as a splint. 'I've done the same thing myself,' he said wryly. I took this to mean he had kicked the television in anger too; I wasn't the only monster around.

That afternoon, Anna and I talked. I told her how much of a strain it was dealing with financial matters. When I went into the office, I said, I was in a war zone. Life was so focused on finances: paperwork that carried little real meaning but had to be dealt with nevertheless. The creditors harassed us, insistent on repayment plans, and hit us with penalty fees and default interest — as if we were naughty children and a good slap was all that was needed to get us to behave properly.

I told her that I'd had a shouting match on the phone with a lender the day before. The woman I had spoken to from the bank had kept hitting us with penalty interest and late-payment fees, even though she knew we had the property for sale and were hungry for any offer we could get, and that the mortgage would be paid out once we sold. She was aware of the parlous state of the property market. Finally we had got our first buyer, and they had paid a holding deposit. The previous day, at the last minute, the purchaser's bank withdrew their finance approval — a symptom of the nervous credit environment. So the sale had fallen over. The woman from our bank said that we would be up for more fees. She was so heartless, callous in her disregard for our situation.

'They hassle and hassle. They don't give an inch. Complete bastards!' I said to Anna. Tears welled up when I told her how humiliating it was for me to front up to other health professionals and admit that I was not coping.

Anna appeared to understand, but she was more concerned that the children were scared after my outburst last night — Emma had cried afterwards. She thought I should speak with them. 'It's better that we see through this difficult time together, rather than separate,' she added.

This was the first time in a while she had spoken clearly about our future. I was buoyed that she wanted us to stay together; I assumed she meant that we could still work things out.

When the girls came home after school, I sat down with all three of them; Ashley was just back from camp. 'Girls, I'm sorry I got mad last night. It's not you. We're having difficult times with money at the moment and Daddy is trying to sell our houses so we can get right again. I wasn't really mad at you.'

They looked relieved. I saw that my stress had been affecting them, and I felt terrible.

I HAD RESTARTED my sessions with Wayne. In my next session, I asked him why I had erupted like this. I hated myself for yelling at the girls and Anna. It had shocked me as well as the family.

'We often take our anger out on those closest to us,' he said. 'They may be the target, but not the real threat. You can't undo what you've done. Anna had the right idea: check in with the children, help them to feel safe, and express regret for your actions. And forgive yourself for being human. Do what you can to make it less likely to happen again.'

I wondered if being male was part of it. When Anna shouted at the girls, they didn't jump as they did when I yelled. Was I more threatening because of my maleness? Did I speak more sharply than her? Was it because I had a deeper voice or was physically stronger? Perhaps it was because I'm a quiet person; when I shout, it's more of a shock. I couldn't tell how fearsome

I appeared to the children, and to Anna. I had never hurt them and couldn't imagine ever doing so.

I'd always thought that my children would not be afraid of me; that they would always feel able to come to me. Yet I'd noticed they went more to Anna than to me if they wanted comforting. I was not as connected to the girls as I once was, as though there was grit in our relationship mechanism.

LATER THAT MONTH, we were unable to settle on the purchase of a block of vacant land by the beach I had committed to buy the previous year, with the idea of building a holiday home for the extended family. The developer's lawyers wrote letters demanding payment and threatening litigation. I had always been able to repay debts, but now I felt ashamed and powerless.

What could we do? Our family solicitor was having a personal crisis. Anna made contact with a new solicitor. At his request, we completed a statement of financial position. We met with him soon after. He sat at his desk, holding a piece of paper with handwritten notes before him, his elbows resting on the table. It was like being in the presence of a headmaster, awaiting disciplinary action.

Eventually he looked over his black-rimmed glasses and said, 'You are in a very serious position. Very serious indeed.' And, as if this wasn't clear, 'It doesn't look good for you.' The creases in his neck came alive with each statement.

I wanted him to stop saying how serious our situation was; it made my sinking feeling sink further. I wasn't sure I'd be able to stay in his office if he kept this up. He outlined the worst-case scenario — eventual bankruptcy — and the best-case scenario — 'You might just scrape through and be left with nothing.'

He thought that bankruptcy was worth considering: it would make the stress of dealing with creditors go away immediately, and then we could make a fresh start. I sensed a human response through his gravitas, as if he had known loss as well. He appeared to have a handle on our situation, and Anna trusted him, so we engaged him to develop a strategy to deal with each potential litigation threat. We asked him to communicate with the developer's lawyers immediately, with the aim of seeking a negotiated settlement over the block of land: a settlement that took account of our real financial position.

After this meeting, I nicknamed him: Doom and Gloom. I came to dread going into his office, because it always seemed to involve bad news.

Over the coming weeks, Doom and Gloom's letters were met by a wall of indifference. A battle loomed.

Anna's property-investing relative recommended a barrister, whom we engaged on the assurance that he'd get us 'out of this mess'. He turned up at our house driving a late-model Mercedes, bringing with him a manner and a business suit that announced, *Don't worry, I'm in control now. Leave it to me*. After a to-ing and fro-ing of letters and telephone calls between him and the developer over two weeks, he was finally told, 'We're going to make an example of this bloke' — meaning that Anna and I would be sued, with the prospect of bankruptcy our only prize.

Bankruptcy.

As our financial security had deteriorated, I'd come to envy those still in work, with children whose educational opportunities remained on track. There would be no option for our children outside of the local high school, which had a mixed reputation, and we wouldn't be upgrading our decade-old cars. The plain-labelled, generic brands in the supermarket became attractive. When I received Doom and Gloom's first bill, I realised that his fee for reading an email was almost the same as one term of weekly dance classes for one daughter. With two girls who loved dance, I queried his bill and put off paying it so that they could continue their classes. It seemed that when you were down, the kicks in the guts piled up.

The Monday night after the barrister told me we would be sued, Pandora's box opened. My sleep was infiltrated by a parade of horrors: the first psychopathic murderer I'd encountered; the alcoholic client who didn't turn up for his counselling session and stabbed his wife in the bathtub; the paramedic I treated after he was called out to a road accident to find the charred bodies of a mother and a child. There I was, conducting an assessment of a father with a history of violence in the same room where a community worker had been stabbed weeks before. Here was my mother's psychiatrist colleague, shot dead by one of his patients. There was the child with a developmental disability who had climbed into a bathtub of hot water, suffering third-degree burns, due to his parents' negligence. The survivors of sexual abuse and violence, the accident victims, the victims of crime — on and on it went.

The next morning, I was a ball of agitation, memories trapped in my head like spiders in a collector's jar. My mind had become a prison of horrors. I could understand why people with psychotic delusions were so tortured: how do you escape a prison that is stuck on your shoulders? I wanted to bang my head against a wall to knock the toxic memories out. It was only when I wrote out descriptions of each memory in my journal, pinning them down onto the page, that I got some release.

Over the past week, a photograph of me as a boy had kept cropping up in my mind. In it, I was wearing shorts, a checked shirt, and sandals. Like a hurdler, my legs scissored over my grandma's low fence, my arms pumping, a gleeful, triumphant expression on my face. I had *always* jumped over grandma's fence — I'd never walked through the dinky little gate. I wanted to go back to that carefree boy and say, 'I'm sorry; I messed up. I'll find a way for you to avoid this pain.' But I couldn't; that boy was now me.

That night, Nick and I caught up to rehearse a Bach duet we'd started to learn. We set ourselves up in his family's music room, surrounded by a piano and guitar cases, sheet music splayed on the sofa. I was still feeling raw. Yet when he closed the door, I was cocooned from the outside world; now my world was black dots on a page, and the sound of our two melodies.

Although the notes in the Bach piece were not exacting, timing was crucial, and mine felt off. It was as if I was being held back by something, as though a hand was dragging on my shoulder. 'Sorry, Nick, my timing's out tonight,' I said.

'Oh no, you're sounding fine,' he said.

I wasn't sure if he was being kind or if I was imagining that I was off. I supposed, given the last two days, I was just out of sorts, and a good night's sleep would see me right in the morning. I looked forward to that.

DIAGNOSIS

8

ON FRIDAY MORNING, the day after I am discharged from Lismore Hospital, Anna and I arrive at Seaview Psychiatric Clinic. It's warm for winter; I take off my jacket. We sit down in a quiet, carpeted corner beside the reception desk. The walls are beige, the ceilings white. A water cooler stands by the far wall. Two others are with us in the waiting area, their heads down, beside relatives, with packed bags in clumps around them. I too have a bag, and a guitar.

Anna is wearing her business lipstick: the deep crimson. Her wonderful smile is hard to imagine now, her face is so taut. We sit in silence.

There are four doors leading off the waiting area. On each, above the word PSYCHIATRIST, is a doctor's name. With the location of their rooms at the bow of this elongated building, these psychiatrists are like captains of the ship. Eventually the door nearest us opens, and the long, familiar figure of Doctor Banister appears. He's wearing his trademark tweed jacket with elbow patches. 'Come in,' he says, looking concerned.

We follow him in. He sits at his desk, facing away from us. Taking a seat on the two chairs behind him, we have a view of his back as he punches away at a laptop. On his desk is a crammed concertina file and the laptop, but little else. It looks as if Doctor Banister has just moved in or is in the process of moving out, even though I know that he's been working here for a long time.

After a few moments, he swivels in his chair to address us, speaking to Anna. 'Tell me what happened … before you took David to hospital?'

She begins her account. Almost immediately, it becomes clear that I am superfluous to their conversation. That's okay with me. The rise and fall of their voices becomes the sound of summer insects — an aural backdrop as I explore the corners of the room. I notice that it has five unequal sides, like a warped pentagon. I'm still not quite sure if I'm dreaming, or if what's happening is real. I might be Alice, and a white rabbit, or something else very unexpected, will appear at any moment.

I see the unframed paintings: originals on canvas. None is hanging; instead, they rest on the floor, propped up by the walls. I'm taken by the large portrait of a young woman. She looks sad, with a long face, small lips, a pointy chin, and dark pools for eyes. Deep red and ochre filters down her face, to her shoulders.

Why are all these paintings here?

Doctor Banister asks me a question, yanking me back to reality. 'David, what do you think brought this on?'

I really am in a psychiatric establishment.

It's an effort to find an answer to his question. 'I … I was very anxious. A barrister came to speak with us. He said that the developer wants to sue us. After he left, I had a huge panic attack; I couldn't stop the trauma memories from coming back. I think it was too much.'

He nods, as though this makes sense. 'Anna tells me that you have been stressed about your property investments. How many do you have?'

Somehow, his interest in this question feels directed more towards our investments than me. Still, I try and tally up the properties in my mind. The market has picked up recently, and some have been selling, at last. But to answer his question, I need to picture each property, remember if it has been sold, and hold the number I have counted to while I picture the next property. The numbers disappear into the fog and won't come out again. 'I don't know,' I say. 'I need to see it on paper ...'

He looks irritated with this answer, but sighs in acceptance.

'I'd like to walk to the beach and go for a swim,' I say.

He chuckles. 'I don't think it would be responsible of us to let you go wandering off on your own. In your state, you'd forget your way back. Best to stay in the clinic over the weekend, and we can reassess things on Monday.'

I'll go stir-crazy if I don't go out. But I don't say anything. I'm not capable of challenging him. The phone rings and he answers; he says it's an urgent call-out and he has to wrap things up. He ends our meeting by telling us that he will be away for several weeks, so I'll see another psychiatrist.

As we re-enter the reception area, a nurse asks to show me to my room. We walk along a wide corridor, with wooden railings attached to the walls at hip height. Coming off the corridor are rooms with large doors. The place has more the feel of a hospital than an outpatient clinic.

My room has two beds, and the nurse points to the one closest to the window. Through the glass, I see a low concrete wall, with a grassy slope rising away from it. There's a smell of cigarettes, even though the building is non-smoking. I deposit my bag and guitar by the bed, and we go to look at the rest of the facility.

When we come back a few minutes later, a young man with John Lennon glasses and a slim, athletic build is lying on the other bed. His hands are clasped behind his head as he stares at the ceiling. I smile in his direction, but he doesn't acknowledge us.

Anna and I hug briefly and kiss each other's cheeks in parting. As she gathers her things, I think of the weight she will be carrying while I'm in here. She's a truly capable and loving person; she takes on whatever problem comes along and deals with it. 'Take care,' I say.

She nods and disappears.

I sit on my bed to face the man. 'Hello.'

'Hi,' he replies.

'I'm your new roommate,' I say, attempting to sound cheerful.

'Okay.'

'How long have you been here?'

'About a week.'

'How's it going for you?'

With an effort, he props himself up on his right elbow and faces me. He's unshaven, dark hair unkempt, with the smell of tobacco coming from him. 'Ah … I've been in before. I come in every now and then for a recharge — to sleep, and to get back on the meds.'

'Oh, right. What do you do when you're not in here?'

'I busk, juggle — sometimes with the circus. It gets tiring.'

He doesn't ask me anything, so I say, 'I don't know how long I'm in for. It's my first time. My name's David.'

'Simon,' he says.

He lies back down, staring at the ceiling again, his face emotionless. He's clearly going to be a quiet roommate — not someone I can share my experiences with.

I decide to unpack. I grab my allocated chair, which stands at the end of the bed, and place it in the carpeted space between my bed and the window, in front of the cabinet. Here I will sit, I decide — read by the natural light, and meditate. With the curtain between my bed and Simon's partly drawn, I won't be visible from the corridor, giving me a thin veil of privacy.

Next I inspect the ensuite, accessible by a sliding door. It's white and stark with absolutely nowhere to hang anything — towel or clothes — when showering, and no ledge to put a toothbrush, soap, or shampoo. The 'mirror' is a piece of shiny metal bolted to the wall. The shower rose is up high, out of reach. No sharp edges or hooks anywhere — it reminds me that the patients here are teetering on the brink.

White, white, white. I crave some homeliness, some warmth, even if it is only in the form of colour. I already miss the sounds and the mess of home.

THAT EVENING, I'M called to see the psychiatrist on duty. He's young, with a pleasant manner, but he wants to hear my story all over again. Strangely, he suggests that I could return to clinical work once I'm better. It's beyond me to explain why this doesn't feel like a good idea; I'm very tired. However, he doesn't mention any restrictions on leaving. I'm still not sure if I'm allowed to go out for walks while I'm here, but I decide to keep quiet for now.

Afterwards, I retire to my hard bed. Lying there is uncomfortable — physically and mentally. How the hell did I end up here? I was a happy child, raised in a loving family where nothing unusually bad happened. I negotiated the turbulence of adolescence without mishap and became a confident, capable adult.

When I was eleven, our family moved to Sydney so that Mum could complete her psychiatric training. It was a time when the seriously mentally ill were managed in large institutions. For several years we lived within the grounds of the Parramatta

Psychiatric Hospital, sharing facilities with the residents — the swimming pool, the sports oval, and the Friday-night movies. The eccentricities of 'the patients', as we referred to them, didn't trouble us; we enjoyed befriending the more personable of them. The only aspect of hospital life that gave me a chill was riding past the locked men's and women's wards. Mum said that the patients there weren't allowed out because they were 'dangerous' — she didn't specify how. I never thought I would end up in a psych hospital as one of 'the patients'.

During that long first night, my mind runs a home movie of the things gone sour in my life. I built a career over twenty years, but now I can't work. Wayne had reminded me that I had eased the distress of hundreds of people and helped them on their way, but right now it feels as though I've helped very few. I completed a PhD and published in my field of specialisation, but all this seems irrelevant. I made investments that secured our family's financial future, but now we face bankruptcy. I was a devoted father, raising three children, but I now feel alien to them. I made a marriage, but it is strained; I have friends, but feel distant from them. At this moment, it all boils down to a shared room with an uncommunicative stranger. A single bed; a white, antiseptic bathroom with nowhere to hang anything; one chair; my guitar; and a three-drawer cabinet — this is my home for now.

My trustworthy, capable, insightful brain was once my strength. It was always able to save me, to make a plan and push me forward. Now it has gone haywire.

I want to cry — to cry inconsolably — but I can't even do that. I am numb. Nothing is working.

THE NEXT MORNING, I'm woken by the barking laughter of kookaburras — the prelude to dawn. I must have slept after all. As the day outside shows itself, I see it will be sunny, tempered by a cool wind. This thought, and breakfast, brings some optimism; I will make the best of my situation.

It is Saturday. A free day, I am told. A supervised outing is on offer; it seems as if I will have a taste of freedom already. A young and lanky clinical psychologist drives the eight or so of us who are taking part in the minibus. I sit in the front and ask him about his work and where he trained. He works at the clinic on Saturdays, he says, and otherwise he's in private practice. *Did I look and sound like him in my early years?* I'm quick to explain that I have post-traumatic stress disorder from my work. He's polite, but it feels as though there is a line between us — I'm the patient.

We stop at a nearby headland and grab coffees before walking across the road to the lookout. We crane our necks, looking for any sign of migrating humpback whales. If they're out there, they're difficult to spot; the ocean is littered with white collars of foam forged by the determined wind, pushing the swell, and it's hard to see anything else. The patients stand in twos and threes. I gather from their conversation that they are mostly at Seaview for drug and alcohol problems — they must be in the addictions ward. Beside me is a young woman with bleached-blonde hair. We get to talking, and she tells me of her competitive surfing experiences. 'Then I realised I was an alcoholic,' she says. She is going home on the coming weekend, and is nervous about relapsing.

What am I doing here?

BY EARLY SUNDAY morning I need to get out again, and I decide to try that walk. There's hardly anyone about. I go downstairs quietly and walk along the corridor through the addictions ward. There's a rustle in the nurses' station as I pass by, but I don't turn to look. At the end of the building, I push open the fire-escape door — it's the least observable exit from the clinic — and hop down a short set of steps. I go straight to the property's fence line, where I've seen patients walking before, and find a dirt track. No one runs after me or apprehends me; so far, so good.

Beyond Seaview's garden, only a short distance away, is a car park, with compacted gravel and low log railings. Interspersed between the parking bays are paperbarks, like old men, wizened and wise, islands of solidity. If only I could be as stable and unruffled as one of these trees. I stop and press my cheek against the cool, smooth, tissue-like folds of bark. The raised branches hold bunches of elongated leaves that shift in the breeze. When I look up, I see diamonds of sunlight set there; but these sparkles are beyond my reach — like the spark that was once in me.

As I walk on, I stop several times to look back at the hospital, with its tiled roof and walls of tessellated brick. It's a dull building, undecided if it should look like a hospital or a small hotel. But I need to imprint in my mind the view I will see on my return; I don't want Doctor Banister to be right about me losing my way.

I go two blocks down a busy residential street, noting the landmarks I'll see on my way back. At the street's end, a view slaps me in the face: lustrous ocean, rock, horizon, and sand. Elation.

The long beachfront is divided by low-lying rock platforms, jagging out into the sea. I have my bathers on and swimming goggles in my pocket. I'm not feeling strong, so I won't do a big swim, but I need to give my body a sensory cleansing — sterilise it from the contamination of my circumstances.

I plunge in, and the expectant cold wraps around me unsympathetically. But to me the embrace is like that of an old friend. I am in the water: water that could have drifted from the bay where I swim regularly. It connects me with home and with friends.

I don't last in there too long, but as I towel off, I feel washed through.

LATER THAT DAY, an older nurse, a small woman, comes up to me as I stand near the nurses' station in the psych ward, waiting for someone I can ask for toenail scissors. 'David,' she says softly. 'I think I knew your mother. Was she a psychiatrist?' She looks up at me intently.

This is such an unexpected question. I'm not sure if I've heard her correctly. My thinking is still slow. How could she know my mother? Mum's been dead for fourteen years.

The nurse mentions a psychiatric hospital and the ward of which Mum was in charge. Finally I get what she's asking me. 'Yes,' I answer.

'I'm Julia. I worked with your mother for many years. She was a wonderful person.'

Her remark sluices open my heart. I wish Mum were here now; she'd know what to do. I wrap Julia in a hug. 'Thank you,' I whisper. And, momentarily, we seem to swim together in the memory of a treasured person.

'Did you know that Doctor Franklin also worked with your mum?'

I had seen Doctor Franklin's name on one of the psychiatrists' doors. 'No, I didn't.'

Now Mum's presence is here — in this woman, in Doctor Franklin, in this place. It's as if I have blinked and the view I had of the clinic staff as removed, even alien, has changed, transforming them into something almost like family. I tell Julia about my situation. 'I've heard that Doctor Franklin specialises in post-traumatic stress disorder?'

'Yes, he does.'

'Could I see him? Doctor Banister said I would get a new psychiatrist.' I'm not keen to return to the young psychiatrist I saw on Friday.

'I'll speak with him,' she says.

I remember to ask for the scissors, and she brings me nail cutters. She asks if I've been taking part in the group activities. No one has told me about the program of activities — or maybe they did when I first arrived, and I was in my mental fog. In any case, I have been left to my own devices. I don't mind, but I have wondered where most of the patients disappear to during the day. She points to a printed program on the ward noticeboard: now I see what I've been missing.

That evening, I sit on the sofa in the common room and place my sheet music on the coffee table. I've brought my classical guitar in with me; my steel-string would have required singing, and I don't have the confidence for that. Cocooning the guitar's body with my thighs, chest, and arms — holding it fast, like a child does a loved teddy bear — I begin to play. Other patients come into the room, to read or do puzzles, but they seem to sense how important this act is for me, and they keep their distance.

Soon I'm a teenager again, sitting in my bedroom, playing into the night's silence. An ageing hippy type walks in and says, 'Hey man, I heard your playing from my room. You're a real guitarist!' Then he walks out. I smile. Someone thinks I'm worthwhile. It's been a good day after all.

9

THE NEXT DAY, Monday, I decide to attend a morning group on 'the locus of control'. This is undergraduate stuff in my field; I can guess what they'll be covering.

A middle-aged psychology intern presents the session, and he hardly looks up from his notes as he reads. It's a bumbling performance, difficult to watch. Even in my fog — which is now more of a haze — I could do better. When it's over, I wonder what the patients could have possibly got out of it.

After lunch, I attend a group-therapy session with Seaview's head psychologist, Peter. This is an open meeting, which means that membership changes from day to day. Today there are four of us, including Peter, sitting in a circle of chairs. Peter's manner is friendly, with a professional edge. He asks each of us in turn, 'How are you going today?'

After we've responded, Peter turns the group's focus to Mario, the large, middle-aged man sitting opposite me. I hear that Mario, who suffers from depression, has been in the clinic for four weeks and is due to leave at the end of this week. He has just returned from a trial weekend at home.

'I felt flat … scared. It was good to see the kids, but I — I don't think I can go back,' he tells us. The words struggle out of his mouth and quiver in the air. His bottom lip protrudes like a child's; tears are close. He slumps forward.

Suddenly, the familiar panic rises. I want to run out of the room. But I also want to stay and face this, for Mario and for myself. I try to generate a sense of compassion for him, to take myself out of my skin. The panic begins to subside, enough for me to say, 'That's how you feel now, but feelings change. By the end of the week, you may feel differently.' I also offer a few more comments. But it's a strain. I want to fix him, and I'm not sure anymore that my advice is good. So I soon shut up, and sit out the rest of the session dumbly.

When it is over and the others have left, I ask Peter for a private word. 'Sure,' he says, and pushes the door shut.

'Thanks, Peter. Look, I'm a clinical psychologist. I have post-traumatic stress from work — it's a long story. I've been depressed and anxious and got really stressed recently, and now I think I've lost it. I've had a breakdown.'

'I see,' he says.

'I don't think I could sit through another therapy group; it seemed to trigger my anxiety. I almost ran out of the room.'

'Um, yes. Stay away from the therapy groups. Have you tried the topic groups?'

I tell him of my experience with the locus-of-control session.

'Give it another go. It'll help occupy you, and shouldn't trigger your symptoms.'

'Do you think I could see you? It's hard to chat with the other patients because of my background.'

Peter agrees to see me the next morning.

THERE'S BEEN TIMES this week when my mind drifts like flotsam, and I re-experience the calm that so infused me on my first day in hospital. It's a remarkable feeling, as if nothing bothers me, as though I'm stuck in the present moment.

But when Anna calls mid-afternoon and says that the developer has rejected our latest offer for settlement, the dread and panic returns. I have to walk up and down the car park for over an hour before the edge is taken off my agitation.

I've got to find other ways to relax.

Off the main corridor of my ward is a room where the nurses write their notes and store things. I walk in and ask the young nurse there for a relaxation CD and a player. She points to an invisible line on the floor. 'Patients are not allowed to cross that line.'

If I were here in my professional capacity, there wouldn't be such an embargo. But she gives me the CD and a player. I listen to it while lying on my bed.

That evening, I have a long guitar-playing session.

THE NEXT MORNING, I'm woken at six-thirty by a nurse — who, I can tell by her red-and-blue uniform, is not from the hospital. She stands by my bed and opens a suitcase. 'I'm here to take blood,' she says, in a schoolteacher's voice.

I'm in a daze, and it takes a while to register what's happening. 'But I had a full lot of blood tests done at Lismore Hospital,' I say. 'Why do you need more?'

'I'm just doing what I've been told.'

'But that's silly. I don't see the point. I'm not giving any more blood.' I rest my head back on the pillow. I don't like her. Doesn't she know what it's like to be prodded and examined every day?

She steps away from me with a sneer, the skin around her nose tightening, as if I am a piece of talking dung. Then she picks up her case, turns, and leaves.

When I'm up, I tell Julia what happened. She says that she will check on the previous blood results.

After breakfast, I catch up with Peter. He's easy to talk to. 'One way of keeping a lid on your panic is to do controlled breathing,' he says. 'Are you familiar with that?'

'Ah, could you go through it?'

'Breathe in to the diaphragm, counting in … two … three, hold … out, relax… two, relax … three, relax.'

It's as if I am hearing this for the first time, even though I've taught breathing techniques to hundreds of others before. He gives me self-soothing statements that I can say: 'I can start again' and 'Anna and I have each other.'

I have made another human connection in the hospital, and my sense of aloneness diminishes a little more. I resolve to use Peter's techniques; I'm sure they'll help.

I just have time to attend a session on 'reasons for change' before lunch. But as I make my way there, I see Alan. I recognise him immediately. He was an ex-policeman, running his own gardening business, when he came to see me a few years before, for help with anxiety. He would come to our sessions with a small notebook in his top pocket, having written down his worrying thoughts as he went about his jobs during the day. He practised the exercises I gave him. I sensed that he had issues from his police career that he hadn't resolved, but he didn't want to look at these — just at how he was managing day-to-day. He made good progress.

Today he is looking calm — still athletic, the same boyish face. I haven't seen him in the ward before, so I guess he must be attending the clinic for a day program. The fact that he is here probably means he had a relapse. Did I fail him?

Alan sees me across the crowd of people milling in the corridor, waiting to go into the various group-activity rooms, and our eyes lock briefly. It is a blowtorch on my sense of shame. I panic. I'm carrying a red clipboard with notepaper, and I quickly bring it up to my chest — trying to look like a busy staff member. It's an automatic response. I back off around the corner and wait for the noise of the crowd to die out. To stop my chest heaving, I count my breaths, like Peter told me.

When it goes quiet, I peer around the corner: he's gone. I think I know what room he went into, and I hurry past it, thankful that there is no window in the door. I'll need to watch out for other former patients who might know me. This mustn't happen again.

TODAY IS ALSO the day of my consultation with Doctor Franklin, when I'll find out what hope there is for me, and why I'm like this. After lunch, I'm back in the characterless waiting room I sat in with Anna on admission. This time, I'm near the door of Doctor Franklin's office. It's very quiet, as if the carpet is soaking up all the sound.

I can't work out what is happening to me. Will I become a nervous wreck, unable to function ever again? Is it my fault that I've ended up here? If so, I can't trust my judgement anymore. Who, or what, do I hold on to? Julia's words, 'Did you know that Doctor Franklin also worked with your mum?' tumble over and over in my mind, becoming solid and polished. I ache for Doctor Franklin to say something about Mum when we meet; it would be a connection, a consoling hand on the shoulder.

The door opens. A man wearing a black suit and a pale-coloured tie emerges. 'Mr Roland?' he says.

'Yes.'

He motions, directing me into his office. 'Please come in.' His face is inscrutable.

I go into a small, square room, ordered and uncluttered. Framed prints of sailing ships hang on the walls — a nautical theme that's incongruous with the setting.

Doctor Franklin looks to be in his fifties, with a thin, youthful face, and hair in need of a trim. He's wearing nerdy spectacles. He sits down behind a gleaming wooden desk, clear except for a computer off to his left and some notepaper before him. I sit opposite, the desk a barrier between us. Behind him is a low bookcase filled with reference titles. There is only one personal item: a photograph of someone sitting at a camp table, silhouetted by the setting sun. I'd like to know who this person is, but Doctor Franklin's facial expression doesn't invite such frivolous enquiries.

He leans forward, forearms on the desk, silver pen poised. 'We know someone in common, I believe?' he says, without a smile.

'Yes,' I say.

And, as if that subject is now out of the way, he asks, 'Tell me, how are you feeling right now?'

'Oh, well — I'm in a daze. I'm not my usual self … my concentration's not good.'

'Can you tell me what day of the week it is?' he asks, and I answer.

The familiar questions unroll; his gaze is intense and enquiring. He asks about my psychiatric history, work history, drug and alcohol use, medical history, family history, life stresses, suicidal thoughts … I see, upside down, my answers turn into indecipherable squiggles on his notepaper. My experiences are being squeezed into the boxes of his psychiatric protocol. It doesn't seem right, summing up rich and ragged life pieces and squaring them away like this. Yet, I remember, it's what I used to do.

Doctor Franklin doesn't always let me finish answering each question before he moves on. I am a small child, hand firmly gripped, unable to keep up with him, an adult striding on. My brain is hurting. I try to slow down the pace with pauses and comments: 'I'm becoming foggy' or 'Can you say that question again?' It feels relentless.

After the questions stop, he leans back in his chair; his face relaxes. He pushes a blank piece of paper across the desk and proffers a pen. 'I'd like you to draw a bicycle for me,' he says.

I know this test from my neuropsychology training. It's a measure of some aspect of brain function, but I can't remember what exactly, right now. I draw it hurriedly, my right hand strangely clumsy.

He asks me to name in a minute as many things as I can beginning with the letter 's', and then to name as many animals as I can. These are word-fluency tests, and I notice how slowly my words seem to come. He doesn't give much away, so I'm not sure how I'm doing. Finally, I am to repeat the sentence 'Tom and Bill went fishing and caught three brown trout.' With a huge amount of concentration, I pull the words out of the fog. I think I get it right, but Doctor Franklin only nods and says he will ask me to recall it again later. I try to rehearse the sentence in my mind, in preparation.

We talk about medication. He doesn't recommend anti-depressants, but sees no need to stop taking St John's wort. However, he wants me to take Remotiv. 'It is the only formulation that guarantees a correct dose of the active ingredient,' he says, making a note to ask the nursing staff to order it in for me.

Doctor Franklin wants to conduct an EEG to rule out epilepsy, and an MRI to have a closer look at my brain. The first will reveal my brainwave patterns, and the second will reveal something important that he does not elaborate on. 'I want the results of these tests before I can be confident of what is going on,' he concludes.

When I ask him how long I'll need to stay in Seaview, he says he'll review this at our next meeting, and if I'm doing well, I could be discharged as early as Saturday.

Anna would like to visit on Thursday evening, I say, and he agrees to let me have dinner with her at a local restaurant.

After this consultation, I'm not left with much of a warm feeling, and my brain is in a serious fog. But I sense thoroughness in his approach about reaching a diagnosis, and I have some hope of release from this place. I don't anticipate that the new tests will show much. I've had a mental breakdown; what else is there to find out?

SEAVIEW IS UNSETTLING and confining. The place is like a drug, slowly immersing me into an addiction to passivity: the inevitable outcome of things being done to you. The next morning, a pathology nurse comes into my room at a reasonable hour and says, in a pleasant manner, that all the tests were done at Lismore except for the HIV/AIDS test, and so into the arm the syringe must go. I accept my fate.

After this is breakfast. I take my food outside, into the warmth of the sun. The garden is dotted with small melaleuca trees with slender, scented leaves. There are no flowerbeds. The few plastic chairs — some cracked with age — are highly prized.

On the lawn I spot Glenys, a patient in her late fifties. I've met her briefly before. I've also seen her from afar.

In the afternoons, I like to shorten my walk and stop for coffee at one of the cafes along the beachfront. Across the road from where I sit, a grass verge drops away, becoming the yellow sand of the beach. Along the verge are needle-leafed she-oaks and squat palms with shredded, plate-sized leaves, filtering the view of the ocean. Pedestrians cruise the footpath before me. I envy them their normality and their freedom, which they seem to take for granted. The cafe's coffee is usually a little disappointing, but for as long as it lasts, I can be a normal person. For as long as I can stare out to sea, before I trudge back to the hospital, I am free too.

Yesterday I saw Glenys. She was walking on the beach side of the road, a takeaway coffee in hand, jauntily attempting to look at ease with herself. The sight cheered me; I wasn't the only one trying to appear normal.

Today Glenys tells me that the patients in the addictions ward, where she is staying, were reprimanded last night for buying coffee outside the hospital. 'Apparently it goes against the program,' she says. 'I don't see what harm it does.'

There's no such restriction in the general psychiatric ward. Even at this low stratum of society, it seems, there are first- and second-class citizens.

Glenys has a straight-backed posture and an educated way of speaking. She looks fit, with an open, tanned face. In the world outside, she runs a dance school. She smiles easily as she tells me her stories, and grimaces when they reach the inevitable sting. 'I can't seem to stop after one or two drinks. I'm having a good time, and by the end of it, I've made a fool of myself,' she says.

I feel the urge to ask probing questions and give advice, but instead, I listen — one confused human being to another.

I'VE DECIDED TO stick with the non-talking groups; the talking variety is too unsettling for me. I've already missed the art-therapy class, but after breakfast I make it to the weekly yoga class in the big common room downstairs. As we recline over the bolsters placed on the floor and wait to begin, I immediately feel as if I'm in the right place.

The yoga is undemanding, and when I tell the teacher afterwards that I've been a regular practitioner, he says that he scales down the intensity because many of the patients are not up to much exertion. Nevertheless, I wish we could do this every day.

Later that day, I learn that the brain MRI that Doctor Franklin has ordered is to be conducted at an outside radiology facility, but the appointment is not for two weeks. However, a young EEG technician arrives. She brings out a white skullcap, with multiple eyelets to poke electrodes through and a tight elastic band around the rim. The contraption reminds me of pictures from the days of lunatic asylums, which I'd seen in my psychology textbooks. She fastens it onto my head and glues the electrodes to my scalp, attaching the wires. These are gathered together at the back of my neck and run down to a battery source attached to my waist. I will need to wear this for at least twenty-four hours; she will come tomorrow to remove it.

A look in the mirror reveals hideous multicoloured growths sprouting out of my head, like an absurd Einstein hairdo. How am I going to walk around looking like this? I saw a young man on my first day here wearing the same contraption and felt sorry for him.

It's imperative that I go for a walk today, as usual. To disguise my head and hide the wires running down my neck, I pull on a beanie and turn up the collar of my jacket. I don't want to freak anyone out.

Alone on the shore, I can forget about my cap. The waves dump on the slanting beach, sending white foam scurrying in cauliflower-like florets. I am barefoot. I'm hungry for the sensation of wet sand pressing through my toes, and the last gasps of the waves slapping my ankles. I'm doing a walking meditation. I need to be anchored. But daggers of thought skewer my attention — lifting it away from my feet and tossing it about. My mind wrests it back to my feet again. It's a tussle, and early in the week there was no clear winner. But with each walk I've taken since, the feet have gained ascendancy.

I return in time for dinner. The dining room has long, coral-coloured tables with matching vinyl chairs. Floor-to-ceiling windows look out over the garden. A wide photograph, one of those typical images of rows of beach boxes, hangs on the end wall. A sign above the entrance says NO STAFF ALLOWED. There is an openness among the patients: no hierarchies here.

Tonight, an obese man with a walking stick spreads three foil-wrapped butter portions on each half of his bread bun. It is difficult either to watch him or to look away, but I don't say anything.

Events like this keep reminding me that I'm a patient. I realise now that in the past I've felt superior to mental-health sufferers. I would say things like, 'It's nothing to be ashamed of. Everyone, sometime in his or her life, will be affected by mental illness,' and then rattle off some statistic about the condition my client was suffering from to reassure them. Well, I am one of those affected now, and I do feel ashamed. I was complacent in my bubble of competency. Now, wherever I look there's a billboard in my face, bellowing: 'Guess what: you're one of us now, boy!'

Although I'm a voluntary patient, I'm worried that I might slip up somehow — show obvious signs of depression, or even suicidal tendencies — and be sent away to a locked ward where I'll have no say: none of the temporary freedom of walks to the beach or coffees. I need to be vigilant as to how I present myself to staff, if I am to get out of here soon.

PRIOR TO MY mental breakdown, I had begun to wonder what my brain was doing to me. How did it create the panic, the sharp oscillations of mood, the suicidal thoughts, the nightmares, and the hovering sense of dread? On Thursday, I attend a talk given by a visiting neuropsychiatrist on neuroplasticity in the brain.

He is holding up the man who layered butter on his bread as an example of neuroplasticity. The man tells us how he had to learn to walk and talk again after a stroke.

The neuropsychiatrist tells us about the role of the amygdala in triggering anxiety. I ask him if there is a difference between fear and anxiety, and the role of the amygdala in this. 'There is no physiological distinction between them,' he says.

Then I catch up with Doctor Franklin for my second consultation. I feel self-conscious talking to him with my stupid cap on. But he is less demanding this time, and chats with me like a colleague about his recent adoption of self-report questionnaires: the type routinely used by most psychologists. He says that he will repeat the speech and memory tests from the first session, and if I do well — and the results of the EEG tests are normal — he's happy for me to be discharged on Saturday.

I'm still concerned about being made an involuntary patient. I try hard to remember the sentence — *Bill and Tom went fishing and caught three brown trout* — when Doctor Franklin asks me to repeat it straight after him, and then to recall it later. It's a little easier this time. He gives me a questionnaire that I recognise as a measure of depression. I complete it, careful not to overstate any of my symptoms.

Late in the afternoon, the technician returns and takes the skullcap off. I tell her that the night before was terrible. The elastic band around my head was torturous. I could only lie on my back because of the battery pack. And I don't say it, but to make things worse, Simon was farting and snoring noisily; deadly pongs drifted over me like rain clouds. I doubt I slept at all. Dawn, when it came, was a wonderful relief.

As soon as I can, I lay clothes, a towel, and conditioner out on the floor in the white bathroom and wash the glue patches out of my hair, free again.

THAT NIGHT, ANNA and Amelia arrive. We go to the cafe where I've been stopping during my afternoon walks. The atmosphere is cosier after dark, with candles, and plastic screens shutting out the ocean wind.

Anna has good news. Doom and Gloom told her today that the developer is going to accept our latest cash offer — he thinks they're in a desperate financial position themselves. This is incredible. Something is going right for a change; there is some light at last.

Our meal together is lovely. I'm glad to be with my family, and so grateful to Anna for keeping things going while I have been out of circulation. Amelia found a blue dog coat in the hospital car park on the way in, with no one around to ask about it. She's still excited. 'Mummy says I can keep it. I'm going to give it to Boo. It will keep him warm.' Boo is the brown, big-eyed stuffed dog that sleeps with her in bed. Her joy over Boo's new coat touches my heart, giving me greater perspective on what matters.

When I get back to the ward, I learn that Doctor Franklin has agreed for me to be discharged the next day, Friday. But Anna had planned for Saturday, and says that she won't be able to pick me up until late Friday afternoon.

The next day, I'm formally discharged in the morning, and, with my bags packed, am free to do as I please. After lunch, I join a youngish and enthusiastic teacher who conducts a tai chi class outside in the sun. I've never done tai chi before. With our bare feet planted in the grass, he guides us through the movements. I notice how uncoordinated many of the other patients are; their stiffness and gaits remind me of the patients in the psychiatric hospital where Mum worked. But I enjoy the session. It's physical, it's outdoors — and I'm now a free man.

Later that afternoon, eight days after my admission, Anna drives me home. Most patients spend weeks in Seaview; I've done well to get out so soon. Still, although I tried to present as well as I could to the staff, I am aware of my impaired cognitive function: my confusion and slowness in thinking. It doesn't seem to be going away quickly.

That evening, we have a quiet dinner with the children — our first time all together since I went into Lismore Hospital. It's a strange atmosphere, as if there is a bogeyman in the room. Eventually the conversation settles on my last night at home.

'Daddy?' asks Amelia. 'Why did you keep coming into my bedroom during the night? You kept staring at me. You didn't say anything. It made me scared.'

'Dad,' says Emma, 'you were supposed to take me and Tina to camp in the morning. You were acting very strange.'

Ashley is now of an age where she thinks her father is strange most of the time and, as if nothing has happened, doesn't make any comment at all.

'I'm sorry I don't remember any of that day,' I say. 'Daddy is not very well.'

10

I'M SITTING IN a fold-up chair under the pandanus palms in the southern corner of the bay's main beach. It's over a week later — three weeks after my initial hospital admission — and I've been coming here most days.

I seem to have regressed since leaving hospital. There are still financial affairs to sort out, but my thoughts get jumbled, like clothes in a washing machine, after only a few telephone calls or a bit of paperwork. My brain is in a fog most of the time.

Here, the world is simple and doesn't demand things of me. The shushing of the waves and the sticky brush of the sea air soothe me. I watch the faces of the wetsuit-clad surfboarders as they pad to the water's edge, anticipating their first wave. I'm happy to be alone. Wherever I look, a persistent dark cloud hangs over my vision, as if it is about to rain. Everything feels heavy at the moment.

I've been bringing my lunch, a book, and writing materials to the beach. Initially I tried reading *Synaptic Self* by Joseph LeDoux, a neuroscientist, but it was like trudging through thick, deep mud.

I would read a page, and by the time I reached the next, I'd forgotten what the previous one had said. Then Anna gave me Elizabeth Gilbert's *Eat, Pray, Love*. It's easy to follow, and it doesn't matter if I've forgotten some of the earlier details. I like her writing, and am buoyed by her story: I'm not the only one on a path of survival.

It's mid-afternoon and I'm reading about her time in Italy when my phone rings. 'David, it's Doctor Franklin. Anna gave me your mobile number. I'm sorry, but I've got some unfortunate news. We made a mistake. The result of the brain MRI you completed yesterday shows that you've had a stroke.'

'Oh.'

'I'm sorry it wasn't picked up earlier,' he says. 'It's affected the occipital region — the vision area of your brain — with some minor bleeding in other parts. Have you noticed any disturbances in your visual field?'

'I don't think so,' I say. I sense the gravity in his voice, and feel I should respond to it. 'What do I do now?'

'Well, it's been three weeks since the stroke happened. If something else were going to occur, it most likely would have by now. But you should see your GP straightaway. Get some advice.'

'Okay,' I say, and we finish the call.

A stroke. Fancy that.

I go back to reading, settling comfortably into my chair; the idea of seeing the GP slips into my back pocket. But my brain continues processing this information in its own way.

After a while, a niggling thought presses: *Perhaps I should see my GP now, or at least make an appointment? Doctor Franklin did sound serious.*

So I fold up the chair, gather my things, and amble off to the car. As I drive the potholed road back to town, the word 'stroke' rolls around in my mind. It doesn't take on any particular shape, but I know that it's a physical condition, something with medical implications, although I can't think of what they are right now.

Then, a thought comes suddenly to my mind: *If I've had a stroke, and that's physical … I haven't had a mental breakdown.* Relief floods through me. *Fucking fantastic.*

WHEN I ARRIVE at the medical centre, I announce with a smile to Joanne at reception, 'I've had a stroke.' I feel a little important with this news. 'My specialist says I that should see a GP as soon as possible.'

Joanne looks a bit startled. She fusses over her computer and says she can get me in tomorrow, late morning. I thank her and walk out; I feel so much lighter already.

When I get home, I tell Anna I've had a stroke. She already knows because Doctor Franklin spoke with her earlier. 'I was so angry towards you before,' she says. 'I thought the breakdown was brought on by your anxiety and self-absorption. Now there's an organic reason. It's just shit luck. I'm sorry.'

Overnight, I digest the news and think about what it means. By the time I arrive at the medical centre the next day, I'm feeling a little less buoyant: a stroke diagnosis flips the view I had of everything upside down.

I'm right to be concerned. The GP, upon hearing my story, says with a stony face, 'You'll need to exercise vigorously, get your heart rate down to fifty, and lose weight.'

This sounds drastic.

She wants me to see a physician, and recommends Doctor Small. 'He's local and has a lot of experience with strokes. What do you think?'

I have no idea what to think. I know Doctor Small as an acquaintance — his wife is in my swimming group — and he has struck me as a gentle man; he smiles a lot.

The GP rings through to Doctor Small's rooms. He says he'll have a referral for an MRA scan for me to pick up the next morning, Friday. The scan will look at my arteries to see if they are the cause of my stroke. His earliest available appointment is Friday week.

The GP also writes a referral for me to see Doctor Mercer, the ophthalmologist in town. I drop by her rooms on the way home, to make an appointment. We know each other through our teenage daughters, so when she sees me standing at the reception desk, she asks why I'm here. I tell her that I've had a stroke, and she asks if I'm driving.

'Yes,' I say, surprised by the question.

'We'd better check you out then, to see if you're okay to drive,' she says. 'I'll squeeze you in after my next patient.'

Doctor Mercer's assistant takes me into a side room and sits me in front of a large metal box with two black eyepieces protruding from it. A contraption holds my chin in place so that my head doesn't move. I can see a large, illuminated circle filling my field of vision. The assistant gives me an electronic button to click whenever I see a flash of light in the circle.

Over the course of a minute or so, pinpoint flashes of light appear randomly in the circle; each time, I press the button. Sometimes there is a larger gap between the flashes, and I wonder what this means.

After this, Doctor Mercer does a thorough eye check in her consulting room. Aside from the slight shortsightedness I've always had, as well as a little bit of astigmatism, she says that my eyes are fine. Then she prints off the results of the visual-field test and clucks in recognition as she looks over them. 'You see here,' she says, pointing with a pen to the circle on the printout, 'you have lost a quarter of your visual field: right superior quadrantanopia.'

And sure enough, there is a neat quadrant of blackness in the upper right visual fields of each eye, as if someone's taken a large chunk out of a cake. This is why I didn't always see the flashes of light. And it explains the smudginess in my vision, which I thought was a rain cloud — it is my brain's effort to make a complete picture of the world with inadequate information.

So, my eyes are fine, the optic nerves going from the eyes to the brain are fine, but my brain (in the occipital area) is not able to process all the incoming visual sense data.

'You should get your full vision back in no time,' Doctor Mercer says, as if it is the temporary loss of a personal possession. 'You'll be fine to drive,' she adds, 'as long as you're aware that you have that gap. Most of what you need to see when driving is lower down.'

I'm left feeling relieved about something I didn't realise I needed to be relieved about. The consequences of this stroke are growing by the hour. I can try to work out a way of dealing with stressors happening outside my body, but it's frightening to know that mysterious things are happening inside. My brain is seriously damaged.

The next day, I do an internet search on strokes. I find out that following a first-time stroke, one in five people die within a month, and one in three within a year. I'm still within the critical period to have a second stroke. Shit. I could keel over at any moment, or become paralysed, or be unable to speak — be trapped in my body!

Anna's mother, stepfather, and grandmother have come to stay for the weekend, straight after hearing of the diagnosis; it's good of them. That night I tell them I'm worried about having another stroke. They're sympathetic, but they don't seem to understand how scared I am — how I'm trying to stare down what I'm becoming certain is impending doom.

On Sunday, as they're leaving, Anna's stepfather, Gary, comes out to see me. I'm hanging up the laundry on the clothesline in the garden. 'Don't stress,' he says, in an admonishing tone. As if I am causing my condition through worry.

'Don't stress?' I exclaim. 'What a stupid thing to say! I could have another stroke at any moment and drop dead. I know you mean well, Gary, but saying "don't stress" doesn't help.'

He starts, as if I've punched him. 'Just trying to help.'

I know that he's a caring person, and he's usually sensitive in his comments. 'Sorry,' I say, relenting. 'It's a difficult time.' I give him a rushed hug, but it's still awkward as he leaves. I can't seem to get anything right at the moment.

OVER THE WEEKEND, the more I've researched the consequences of my stroke, the more my mind has become flooded with questions. Should I be taking blood thinners or some other kind of medication? Should I be resting or exercising vigorously? Is it all right to read, or will that tax my brain and bring on another stroke? But I'm in limbo until I can see Doctor Small in a week's time.

Come Monday morning, I can't stand it any longer. I decide to call Doctor Small's rooms, and am put through to him. He is surprised when I explain how little I've been told so far; he hadn't realised that I'd been misdiagnosed in the hospital and that, because of this, I haven't been given specific instructions or medication. He says that I should be taking a small dose (up to 300 milligrams) of aspirin daily, that I should rest rather than exercise, and that I most likely would have had another stroke by now, if it was going to happen. I have the MRA scan tomorrow, and on Friday he will review my medication and the MRA results. He doesn't think anything will change for me until then.

My relief is instant.

The next morning, before I go for the MRA scan, I catch up with Doctor Banister, the psychiatrist, for the first time since the stroke diagnosis. It is our third session in his private practice. He has my MRI results — they were sent to Seaview, and Doctor Franklin has passed them on.

As soon as I sit down, Doctor Banister says, 'You've had a major stroke.' He shows me the film: the area of the lesion looks about the size of a golf ball. 'You're looking good, considering this. It's amazing.'

He asks how my memory's been, and about my mood. I'm still conscious of presenting at my best, concerned that I might be made an involuntary patient and locked up. He doesn't mention the misdiagnosis, or his part in it. He asks me how I found it being in the clinic. The staff were caring, I say, the food better than I was expecting, and it was good to have a break from the legal and financial pressures. He looks pleased to hear this.

I'd had my first session with Doctor Banister several weeks before my hospital admission. I thought it made sense to have a psychiatrist advocate for me with my insurer, because my policy required feedback from a medical doctor, and my injury was psychological. Also, I thought that a psychiatrist might have a new viewpoint on my treatment, as psychiatrists come from a more biological perspective than psychologists. Ian and I had agreed that we didn't want him to take on this role; he had recommended Doctor Banister as someone who was a strong report writer.

My initial impression of the doctor had been positive: he was friendly, smiled, and seemed concerned. But during his assessment, after each answer I gave, he swivelled his chair around and tapped into a laptop on his desk, while I sat, waiting, on his sofa. It was like being interviewed by a reporter, not someone who was developing a therapeutic relationship with me. When his questions were over, he pronounced, 'You've had major depression. You should've been put on antidepressant medication straightaway.' He thought the treatment I'd received to date — the psychotherapy with Wayne and the St John's wort — was inadequate (even though he hadn't sought much information about Wayne's treatment approach). 'If you'd taken an SSRI, you wouldn't be here speaking with me now,' he continued. It felt like a blunt and hasty conclusion. How could he be so certain?

I knew that SSRIs did not work for everyone, and they had a high placebo effect. By knowing these things, would the placebo effect have been mitigated for me?

Doctor Banister didn't think I had major depression anymore, so he said antidepressants weren't necessary. While he saw no merit in me taking St John's wort, he didn't suggest that I should stop.

He thought that my financial pressures were the major cause of my symptoms, not my psychology work. 'But what about the trauma stuff: the work in jails, the victims' compensation clients, the Children's Court? I dropped in a report from Wayne to your secretary a few weeks ago about this?' I said. He seemed to be missing the whole trauma story.

'Oh, did you? I haven't seen this yet.'

I had left the session feeling confused: it felt as if I'd done something wrong. He'd challenged Wayne's diagnosis of post-traumatic stress, hinted that I should not have taken my own initiative in regard to medication, and implied that the help I'd received from Wayne was — what, useless?

In our second session, two weeks later, not long before my stroke, we'd discussed the trauma link further — by then he'd read Wayne's report. I mentioned that some of my triggers were stories of health practitioners being harmed or even killed by their patients; in particular, there'd been a recent local incident of a mental-health worker being murdered by his patient, and this had triggered memories for me. After this discussion, Doctor Banister seemed more accepting of the significance of my trauma.

Today, as he turns away from me at the end of our session and leans over his desk, he says: 'If you'd been taking antidepressants, you wouldn't have had the stroke.'

What? How does that work? I'm too flabbergasted to respond, and he doesn't elaborate.

As I return to my car afterwards, I'm angry. *So, I'm wrong again,* I think. *I caused my stroke by doing what I thought was right and following others' advice. Terrific.*

THAT NIGHT AT home, I pull the CD of the MRI images out from the radiography envelope. Once it is inserted into my computer, I watch the ghostly images of my brain as I click on different views. They're both beautiful and unsettling. Here is the wonder that makes me human, and here is the damnable thing that gives me nightmares, panic, and confusion. It's a love–hate relationship with my brain at the moment.

FRIDAY ARRIVES, AND I find myself sitting in Doctor Small's waiting room. He comes in. 'David, good to see you, although I guess not in these circumstances.' He extends his right arm and we shake hands; his handshake is firm. I follow him down the corridor to his office.

The practice is in an historic, impeccably renovated house. His room is spacious, and, while obviously medical, almost homely. On a shelf behind him, I see a large photograph of his son, whom I have met, playing football; there is a footy on the mantelpiece over the fireplace. Through the large bay window — the lower part obscured by white lace curtains — I see a hedge. While it's bright outside, it is pleasantly cool inside.

He shuffles through the MRI and MRA scans in an unhurried

manner, every now and then placing one up on the light box beside his desk. I notice his broad shoulders and short, greying hair. He has a military-style moustache, which, I think, makes him look distinguished rather than unfashionable.

He squints, examining the scan, and points his finger to show me how my left cerebral artery, at the back of my head, caused the stroke with a blockage: 'an infarct', he calls it. I have an area of damage extending from the left occipital lobe into the left hippocampus. This region borders on the left temporal lobe. The occipital lobe processes vision and images, the hippocampus processes everyday memory and factual memories, and the temporal lobe processes sound and speech, and long-term memory.

'Your arteries are fine,' he says. 'There are a few more investigations we can do, but my guess is that we won't find anything, and your condition will remain idiopathic: a fancy medical term for "we don't know".' He explains that it could have been a random blood clot from anywhere that found its way into the cerebral artery. I don't have the usual risk factors for a stroke: no family history of vascular events, no hypertension, normal lipids, and I'm not diabetic. I don't smoke and I have a modest alcohol intake. At fifty-one years, I am a 'young stroke'.

He will arrange for an echocardiogram of my heart, to exclude sources of embolism; a carotid Doppler study, to check my carotid arteries; blood tests to look for clotting, inflammatory markers, and homocysteine; and a full lipid screen.

When I mention Doctor Banister's assertion that I wouldn't have had the stroke if I'd been on antidepressants, Doctor Small responds with a look of bemusement.

Since I've been out of hospital and gone back to normal life, I've been buckling under daily demands that, before the stroke, were manageable. In Seaview, I only needed to look after myself: there were no chores to be done, and I could rest and go to quiet places whenever I needed to. Now I get incredibly tired, and often sleep during the day. I tell Doctor Small how fatigued I get with general walking and light gardening, and how it takes me the rest of the day to recover.

I explain that I've forgotten a lot of things, too, such as general facts and names (of actors, writers, musicians, song titles, movie titles). It's different, I say, from having a word on the tip of the tongue, one that you know will come in time: some words have completely gone. It's as if they've dropped off the back of a truck, scattering along the roadside, and I've kept on going. I've forgotten the names of acquaintances, although I remember the names of close friends and family — but even so, I now have to think about these, to recall them, in a way that I didn't need to before. Conversations are tiring. In the middle of speaking, I lose track of the point I'm trying to make.

I also have an almost constant low-grade headache, I tell him. When I drink a glass of wine, the hangover-like effect lasts for a day or two after. I like a glass of wine now and then, and the occasional beer, but the consequences are not worth the pleasure of a drink.

One good thing, I tell him, is that I can sing and play music as I did before. It's harder to remember the words of songs, but hitting the right notes is the same.

Doctor Small says that there's not a lot I can do. He doesn't think that I need to take blood-thinning medication because of the condition of my arteries, but recommends 100 milligrams of aspirin daily as a preventive measure. I am to take it easy and avoid stress as much as possible.

'What about cognitively?' I ask.

'Read, but nothing harder than the newspaper,' he says.

Good. I should be fine with *Eat, Pray, Love* then.

TWO WEEKS LATER, Doctor Small has the latest test results. The bloods are all normal. I have no inflammation of the arteries, no heart condition, and no problem with the carotid arteries. It's good news that I don't have any of the nasties, but the cause remains uncertain — so there are no pointers to what might happen in the future.

The only physical sign that Doctor Small finds is low blood pressure: it's one hundred over sixty-five. He wonders if how easily I tire with minimal physical activity is because of damage to the cardiovascular centres in the upper brain stem. The MRI, he says, is not going to show everything. 'It's unusual for a doctor to recommend this, but I'd encourage you to have more salt. And keep up your fluids.' He tells me to walk only on flat ground — no steep inclines — for the time being, and to avoid swimming. I'm doing a Pilates session once a week, and he thinks this is okay.

The risk of another stroke is low, he tells me. 'Time is your best friend.'

'What about stress?' I ask. I've told him of the financial pressure we've been under, and about my anxiety attacks. 'Could this be a cause?'

'I couldn't stand up in a courtroom and say that your stroke was caused by stress; no doctor is going to do that. The medical science behind that is unclear. Personally, I believe that stress can cause a stroke.' And he tells me of a close friend who he said had worried himself into a stroke.

There's one other thing I'm still puzzled by — the misdiagnosis. I know that Doctor Small ran a stroke unit in Melbourne before moving to our area, so he might have some idea of what the doctors were thinking. I lean forward. 'Why did the CT scan at the hospital come up as negative?'

He says that it's normal practice to order a CT scan immediately if a stroke or a transient ischaemic attack (a 'mini-stroke') is suspected. But a negative CT scan does not rule out stroke: the damaged area of the brain can appear normal soon after onset; the stroke region may be too small to be seen on a CT scan. An MRI is more accurate, but it takes longer and is harder to arrange. 'Diagnosis is a matter of probabilities. Your only risk factors were being a fifty-one-year-old male and suffering from post-traumatic stress disorder. A stroke for someone like you is a far more likely outcome than a fugue state, which is rare. I often tell my medical students, "If you hear hoof beats in Texas, it's unlikely to be zebras."'

He explains that a neurologist should have seen me within twenty-four hours, but the hospital doesn't have a neurologist, or an MRI machine. 'In your case, I would've arranged for an MRI at the radiology centre up the hill, and given you a big dose of aspirin as a matter of course.'

I mention that I've read something about a medication that reverses a stroke. 'You mean plasminogen activator?' he says. 'It breaks up blood clots in the arteries of the brain. To work, it needs to be injected within six hours from the onset of symptoms. It was probably too late for you. In the end, I don't think the misdiagnosis has changed the outcome, medically speaking.'

This is reassuring. My stroke happened during the night, and Anna got me to hospital as fast as she could: faster, she reckoned, than if she'd called the ambulance. We did all we could.

THIS STROKE THING is a process of discovery. The invisible hole in my head is a trickster; I don't know when or how it's going to trip me up next. Some days my brain decides to work, and on other days it's like a sullen teenager, refusing to cooperate for no clear reason.

My body's not behaving properly either. In Seaview, it took me several days to get up to a reasonable walking speed, but now, at home, I seem to have regressed. I walk daily into town, a distance of a kilometre, to pick up the post, do a bit of shopping, and stop for coffee at my favourite cafe. Marion is one of the regular waitresses I like to chat with; she is a singer in a reggae band.

One morning, I'm standing at the counter as she takes my order when she says, 'Dave, you're scaring me. You're wobbling from side to side. Come and sit down.' She guides me over to a chair.

After the coffee and a rest, I head home, walking like a frail old man.

What is happening to me?

Over some weeks, I work out which activities bring on dizziness. They often involve physical exertion, such as swimming and long walks. Gardening — especially weeding — and packing and unpacking the dishwasher also bring it on; I think this is because of the up-and-down movement of my head. Pushing the lawnmower brings on extreme fatigue, and I need to lie down afterwards. Physical fatigue also drains my mental concentration, and then everything is harder.

I'm bumping into things on my right side, too. The grip in my right hand is not quite there, and more than once I drop cups onto the kitchen floor. My handwriting is clumsy, as if my right hand is drunk and stumbling across the page.

I can't multitask anymore — or perhaps it's more that I can't filter out distractions. I need to complete one thing before moving on to the next. If I'm derailed from my mental tracks — say, if someone speaks to me or if the phone rings — I have to work out, by a process of detection, what I was doing before the derailment. If this fails, I turn to the next thing that comes to mind, and I forget to return to the first task. When I read professional books, I can usually get the gist of the concept I'm reading about, but once I reach the next idea, a wall goes up, separating what I've just read from what I'm reading next. The ideas don't hang together — they're like a string of beads with the string taken out. I'm left with an attractive but useless pile of beads.

Each morning I write a to-do list, which I carry around with me. This is my best strategy for staying on track — so long as I can remember to take the list with me. But on the days when I wake up already worn out, I stare at the paper, waiting for items to surface in my mind. Nothing comes. Then I have no strategy.

I do most of the grocery shopping, but without a list it's disastrous. Once I'm at the supermarket, I have little memory of what we have at home. If I have made a list, I've often forgotten to take it: so many things divert my attention that it's like going through a mental minefield before I get out the door. So, listless, I cruise the shelves, dropping things into the trolley as if we have nothing at home. I'm always compelled to buy two cans of tinned tomatoes and a tin of kidney beans. Even the kids have noticed the growing profusion of these in the pantry.

Each time I return to the house, I have to place my car keys and wallet precisely in the left corner of the sideboard. But often I forget, and they end up wherever I happened to put them down. When I need them next, I panic, searching the house with absolutely no idea of what I've done with them. If the kids are around, I call out to help me look; they usually find them quickly. Sometimes the keys are on the sideboard, but to the right of the corner. I appreciate how much my brain was coordinating my everyday life — now, I'm like a car without a steering wheel.

When I need to make a decision on the spot, or under a time constraint, I become overwhelmed. I go into freeze mode, like a frightened animal. If someone else is there, I want to run away, my body telling me that this is the only way to escape this feeling of being under siege.

I also freeze when people innocently ask an open-ended question — say, Anna might ask, 'What would you like for dinner?', which is a much harder question than one requiring a yes-or-no response. For me, it's like having to find files in a rusty old cabinet; I have to extract the meaning of the question first, locate the file on dinner options, sift through these, and then make a decision on what I feel like having. I see the confusion on the questioner's face when I can't respond to what seems straightforward. In time, I learn to say, 'I can't answer your question right now' or simply 'You decide'. My brain is working slowly, but the rest of the world is going as fast as it ever was.

Something else has changed. I'm attending a stream of medical appointments and often getting lost. It's ridiculous: I'm disorientated when driving to familiar places and, sometimes, even on the way back home. I don't feel stupid, just … incompetent. When I tell Doctor Small about this, he calls it 'topographical disorientation'. I buy a GPS for the car, and the struggle to learn how to use it is worth the effort. One less thing I need to rely on my brain for.

I work out that I have three levels of brain incapacity. The first is 'fog brain'. It comes on like a mist descending. I'm unable to understand what's just been said to me, or what's being asked of me, or how to do something. I feel like a child around adults who are making decisions about me — powerless to influence the outcome. Financial and legal matters, in particular, bring on fog brain. In these situations, I often have to withdraw, or explain that I simply don't understand.

It's fog brain that makes me decide to start sending out a regular email to update friends, family, and acquaintances on my recovery. When someone asks me, 'How are you going?' with that concerned look, I'm flummoxed. They don't realise how exhausting it is to explain. I can't tease out the most important thing to say, what I may have told them before, and if I would be repeating myself (as I often do). Usually, I just reel off the latest test results and my latest symptoms. So I make up an email list of those I think would be interested, and add any person who enquires about my health.

As my thinking starts to improve, I begin to really enjoy writing these updates. I can create for myself a little virtual world, pretending that nothing bad is happening outside of my stroke, and let my sense of humour creep into the words. I refer back to earlier emails to avoid repetition and to make it a continuing story. Some of my readers, when I meet them in person, tell me that they look forward to my updates or that they feel they're on the journey with me.

The second level of brain incapacity, 'rubber brain', occurs when I've been concentrating too long: the conversation has gone on for more than thirty or so minutes, or the noises around me have sucked my mental energy dry, or I'm speaking to a new person who wants to know all the details of my story. With rubber brain, when someone speaks to me, I have the sensation of his or her words bouncing off my brain: nothing comes in and nothing goes out. It's absolutely time to stop and recuperate. I often think that it'd be startling for these people if they could see how their words seem to ping off my brain.

The worst level is 'sore brain'. I first discovered this stage a month after the stroke. I had been invited to an engagement party for two of my swimming buddies, James and Phillippa. I wasn't keen on going because conversations with new people were by far the most tiring. And there would be a crowd and music. But I went because they're good friends.

Lily gave me a lift, and we arrived early. I talked with the few others there. The music was low, no more than background noise. I thought, *This is going all right.*

But half an hour later, the house was full of people. I ended up squashed in the kitchen, holding a glass of sparkling water, talking with a young lawyer. I hadn't told her about my stroke; I wanted to see how I'd go being normal. We'd been talking loudly to be heard over the din.

Then she disagreed with something I'd said. Her eyes were fixed on me: she meant to get her point across. It was at that moment the change happened: suddenly, I couldn't understand what she was saying; she may as well have been speaking a foreign language. Her words, the conversations around me, and the music were like pins or darts in my brain: my head hurt, not as it does when I have a headache, but as if I was being stabbed. My skull felt too small for my brain, as if it were trying to get out.

I had to leave. I told the lawyer that I needed fresh air; she gave me a disdainful look in response. I pushed along the corridor, through the bodies jabbering like cockatoos, to the front door, and burst out into the night air. The noise from the house exploded outwards behind me, like a massive fart.

I walked a few houses' length along the road; it was a quiet cul-de-sac. This was better, but my brain still hurt. I looked up into the clear sky. Ordinarily I enjoyed the stars, but tonight they seemed as coolly distant from me as the old me did — the one who would've jousted with the lawyer, tolerated the music, and thrived on meeting new people.

I wanted to wait until Lily was ready to go. I sat, stood, walked, and paced, wrapping my arms around myself as the cold sank in. The pressure in my brain was unrelenting. *God, this is awful.* Eventually I walked back into the house, found Lily, and told her I wasn't holding up too well. I felt like a killjoy. She said that we'd go soon. I went out and walked up and down some more.

Half an hour of the party: that's all I'd lasted. What a miserable thing a brain injury is.

11

I GATHER THE kids, and we sit on the verandah. It's time for us to have a talk. I've been getting sore brain frequently at home. The household noise and the kids' demands, on top of my usual activities, have been tipping me over the edge.

'Girls, you know how I've had a stroke and my brain's not working like it used to?' I say.

They nod like this is old news.

'Well, when I'm really tired I get a sore brain.'

The younger two laugh. Ashley is inspecting her nails.

'My brain really hurts, like I have a bad headache, when there is lots of noise or when I've been working. If I say I've got a sore brain, I need you to stop asking me for things and leave me alone for a while. So I can sleep or go for a walk by myself.'

'Oka-ay,' the younger ones say.

Looking at Ashley, I say, 'Did you get that?'

'When you've got a sore brain, we have to leave you alone,' she recites. She glances at me with that expression: the type perfected by teenage girls, which demotes the parent to the status of a worm.

After that, the girls do try to leave me alone, when I tell them that I have sore brain. I don't always get enough time to fully recuperate; the sore brain doesn't always go by the next morning, leaving a hangover-like trace. But it is a help.

Yet I don't think Anna completely understands that I can't meet the usual family demands as I used to. After all, she has relied on me, as I have on her. When I've got sore brain and she wants a decision from me, or we need to work something out, all I can say is: 'My brain's not working.' Often then I see a look on her face that is a mix of frustration and doubt. Once, she says: 'When *will* it be working again?'

One Thursday, six weeks after the stroke, Anna calls me to join her for lunch at a cafe. This is nice; we haven't done it for a while. We sit outdoors under a leafless poinciana tree, which is letting through the lukewarm winter sun.

After ordering, she says, 'Dave, I'd like to talk about how we do the separation.' It is said in a matter-of-fact way, as if we'd agreed that this was on the agenda for discussion. She waits for my response. But I'm dumbstruck.

Eventually I say, 'I can't talk about this right now. I thought we were going to review everything at the end of the year ... after all the financial stuff was out of the way?'

She looks taken aback, as if I've pulled a rabbit out of a hat, and it stalls the conversation.

When we'd started the relationship counselling, the agreement was that we would commit to staying together for six months while we worked on our marriage. After Anna returned from overseas and these sessions finished, she didn't raise the issue of separation again — or if she did, I'd been too preoccupied to register it.

In my early days of seeing Wayne, she would often ask, 'Did you speak to Wayne about us? About the family?'

'No,' I would say, or 'A little.' Once I elaborated: 'Anna, I'm just trying to survive. Getting through the day is hard enough for me at the moment. Most of what we talk about is to do with this.'

I'd felt then that she didn't understand how much post-traumatic stress and depression had taken over my mind, how difficult it was for me to think outside of this. I suggested that she speak to Wayne; this would help her to get a better sense of what I was going through, and let her get things off her chest. I suspected, too, that she wasn't sure about the quality of advice I was getting, and I wanted to reassure her.

At the following session, Anna spoke with Wayne while I waited in reception. She was inside for a long time, and when she came out, I asked how she'd found talking with him. 'He's the first person to really listen to me,' she said. 'He understands what I'm going through.'

Anna felt reassured that I was seeing a capable professional. I thought that she was a little more patient with me after this, although she still asked from time to time if I'd discussed our relationship issues with Wayne. I sensed that she was hurting, but when our financial problems began to close in, it became even harder for me to think about the relationship. Mentally, I put our marriage issues on the back burner — hoping I could get to them when we'd dealt with the finances, and I had the mental space for it.

I wasn't totally oblivious: I could see that Anna was socialising more without me, but I didn't think she would really leave. I thought she'd stick it out, like me, until we could give our relationship proper attention. I'd told Anna before that whenever the notion of separation came up, I went into freefall. It was like the plug was being pulled from the sink, and I would drain away.

At the cafe, the food arrives. We talk about safe subjects as we eat and then make our individual ways home.

As I drive, I feel a deep sadness for what is happening to my family: we are coming apart. Is it possible for things to get any worse? The chasm between Anna and I yawns wider and wider; it feels as if I'm falling into nothingness.

AS IT HAPPENS, my old mate from Sydney, David, arrives the next day to stay for the weekend. When he learnt of my stroke, David arranged to visit as soon as he could get away. He and I first met while working in the prison system when we were both young psychologists. We discovered common interests (music and fitness) and we enjoyed cracking jokes with each other. Our lives have paralleled since then: marriage, children, and private practice. I'm excited to see him.

David's presence is a morale booster. Anna trusts him — she has gotten to know him and his wife well over the years, and our families have even holidayed together a few times. I mention to David that Anna has talked about separating. He's concerned. He's a skilled psychotherapist, and agrees to speak with us individually. That night he speaks with Anna.

David's become a keen photographer, and on Saturday morning, through a chance meeting, we tee up a photo shoot with a local band who want new publicity shots; he'll take their pictures when they play at the markets on Sunday morning. For most of Saturday we spend time at the beach, walking and talking. I don't feel so alone.

On Sunday, before he leaves, the three of have lunch together and he recommends one last try because, he says, 'of all the good things you have: your relationship, the children, your mutual friendships, your finances'. 'You've got to earn your separation,' he tells us. He encourages us to have another go at therapy. 'If you can't do it yourself, get some help. You've got to try and tease out anything that might be improved.'

His words clear the air between us, and it feels as though Anna softens somewhat. I'm reminded of how much we have to lose if we don't work things out.

As I drive David to the airport, he says, 'She loves you enough to give it another chance. I don't think it has gone past the point of no return. But you have to come back to life; you can't expect Anna to lead a life of duty to you. You have to do everything in your power to be that person she married.'

The next day, Anna and I speak about having a 'last go'. We make a plan to follow up contacts that could resolve our financial and legal predicament.

Anna makes an appointment with a sex therapist. Part of the condition for the therapy is that we remain together for three months while we work on things. It feels like the last roll of the dice.

SOME WEEKS LATER I'm back doing battle with creditors and lawyers, and by Friday, I'm wasted. I go for a gentle swim, get a haircut, and decide to treat myself to a movie.

I go to the cinema. When I survey what's on offer, there's not a great choice. In the end I fork out for *Disgrace*, starring John Malkovich. The write-up says that it's about an English professor who seduces one of his students, resigns in disgrace, and then recovers from his fall. I'm not drawn to the reason for his disgrace, but the idea of someone messing up, recovering, and then redeeming themselves is appealing in my current circumstances.

After his fall from grace, John Malkovich's character, David, goes to live with his daughter, Lucy, on a farm in post-apartheid South Africa. While there, young black men invade the house and rape Lucy. David is pushed into a toilet cubicle and doused with alcohol; a lighted match is thrown in and the door locked. He frantically puts out the fire with the water from the toilet.

Afterwards, this scene keeps popping into my head. It's as if I was there as it happened. That night, I re-experience memories of burn victims to whom I've provided therapy. I remember the manager of a supermarket that caught fire, an accident for which he felt responsible. He subsequently became obsessed with fire tragedies and couldn't stop watching news reports of them.

For the rest of the weekend, suicidal thoughts invade — they're strong enough to really scare me. On Sunday night, I tell Anna what's happened. She says it is awful that I feel this way, and reminds me that I have three beautiful daughters. She confesses that she's also angry and sometimes wants to tell me to snap out of it.

When I speak with Wayne several days later, he reminds me not to watch the news or movies that are emotionally charged. He's right, of course, but movies that I think are going to be all right, such as *Disgrace,* ambush me. Perhaps I need to be like the alcoholic who decides on a blanket avoidance of any occasion where alcohol might be involved.

But I still need diversions. My preference becomes comedies or human-interest stories based on true events. I avoid violent, highly sexual, and despondent movies. We borrow DVDs from our local video store each week. After the incident with *Disgrace*, I ask the store manager, Jarrod, for his feedback before I hire a movie. If either of us is uncertain about the content, I don't get it. I become my own censor, with Jarrod's help.

I also try to get a lift from doing things that are proactive. My dealings with creditors and lawyers are all reactive: responding to their demands and threats, always on the back foot. I've been editing a manual that Anna has written for one of her business projects, and I've been helping to revamp her website. We've also been talking through an idea she has for an instruction book. These tasks are fatiguing, but I can concentrate for longer than when doing the reactive tasks, and it's much more satisfying —

I feel useful. Growing vegies in the garden also gives me this feeling.

On Sunday mornings, our three daughters participate in Nippers. The new season started a few months after the stroke. In previous years, I mostly helped out with Ashley's group, doing water safety for the age manager, Bill. (Age managers organise the water and beach activities for those in their charge, and parents provide assistance to the manager in whatever ways they are able.) Nippers cannot operate without the required number of water-safety people when the children are doing their water events. Water safety is done by those who've completed volunteer surf-lifesaving training and have passed the annual proficiency test, demonstrating water rescues, CPR, and physical fitness.

Four months after the stroke, I return for a Nippers session; I've just recently gone back to swimming in the ocean and the pool, and I'm feeling strong enough to give it a go. Bill sends me out on a yellow safety board, the type used by lifesavers. It is long, wide, and fat, and can carry two people in a rescue situation.

A northerly wind blows as I paddle, and the swell rears up, breaking in my face. I'm pushed back into shore a few times, but I don't give up. Yet with each effort to move through the breakers, I become weaker. I'm puzzled; normally I'd get through this.

I look back to the shore and see suddenly that I've drifted a long way from the group. Bill already has another water-safety person out in the surf; I've become redundant. I turn the board around and edge it back into shore, catch a final wave, and cut between two groups of children, their capped heads bobbing in the surf.

As I pull the board onto the sand, one of the Nippers officials comes up to me. 'You don't have the board skills to handle these conditions,' she says. 'You've already run into the kids on the board once before, haven't you?'

I'm dumbfounded. I stare at her, not knowing what to say.

'Stick to the rescue tube; you're a good swimmer.'

'No, that's not right,' I say finally, as I drag the board further up onto the sand. She is already walking away. I realise how angry I am in her judgement of me — and I haven't run into the kids before!

I catch up to her and say, 'Look, I'm recovering from a stroke.'

'Well, you should know your limitations. We can't have you putting the kids at risk,' she says. There is no sympathy in her face.

'I've only just realised I can't do it,' I say, waving my arm towards the surf. She looks unmoved. Why is it that when you're trying to get back on your feet, someone goes and knees you in the guts?

I jog back to Bill's group. 'Sorry, Bill,' I say when I reach him. 'I've lost strength. I think it's the up-and-down motion I'm finding hard.'

I'd told him before we started that I'd missed the recent sessions because of a stroke, and I wasn't sure how I'd go. He'd been sympathetic. 'Don't worry about it, Dave,' he said. 'You can help out on the sand.' I appreciated this. Bill had had some recent health setbacks himself.

I won't do water safety again. I can't expose myself to put-downs like this while I'm recovering; I'm too fragile. And I certainly don't want to be a safety risk to the children.

AS THE WEEKS go on, I begin to seek out easy-to-understand articles and picture books on brain function. This type of reading is so much easier than the legal documents put in front of me.

The Human Brain Book, by Rita Carter, soon becomes one of my favourite books. I stare for long periods at the sliced images of a real human brain inside its skull. In the diagrams, the anatomically differentiated brain areas are artfully coloured and individually named; it's like going through a natural-history museum, each page of images another room. In my own brain, I try to feel the shapes I see on the page — mental fingers feeling around inside my skull. My brain and I are getting to know each other. As strange as it may sound, the more I can remember these names and shapes, the more it feels that I can have a dialogue with my brain, and the more influence I can have on it.

I start to sense that my brain is not a black box, but a menagerie of characters: sometimes working together, sometimes not. They begin to take on personalities: the amygdala is highly strung, an overanxious protector; the hippocampus is a methodical organiser, the cataloguer of memories; the prefrontal cortex is the master controller — at times over-serious, at times visionary and creative, and often at loggerheads with the amygdala. Then there's the mysterious limbic system: the emotional and impulsive one. The dutiful hypothalamus, the accelerator pedal of the fight–flight response, makes hormonal decisions, together with its compliant cousin, the pituitary gland.

A few weeks after the stroke, an ex-colleague lent me two books: *My Stroke of Insight* by Jill Bolte Taylor and *The Brain That Changes Itself* by Norman Doidge. It's taken me a while to get round to reading them, but I feel ready now.

I'm still often plagued by fog brain when I read, so Taylor's book looks the easier to start with. Soon I am riveted by her account of the early stages of her left-sided stroke, which occurred as she was getting ready for work. The onset of my stroke happened while I was asleep, and I didn't get the extreme motor loss that she did. But she describes the sense of peace and connectedness, and a lack of perception of danger, that she experienced immediately after her stroke — something I also experienced. She loses me a bit when she talks about reaching nirvana and dissolving perceptual boundaries, but until that point I am right there with her.

What I find most comforting in her book, which is the first thing I've read that describes my experience, is in the final section. Taylor lists the things she needed in her recovery. I identify most with her need for sleep; for others to speak slowly and softly, and to tone down their energy levels; for a reduction in stimulation from electronic media; for others to ask specific questions rather than general ones, giving time to answer; and for others to accept her for who she is today.

Next, I begin Doidge's book. It is a revelation, taking me into a curious, fascinating world. And it gives me hope that I can do something about my malfunctioning brain.

As a boy, I traded football cards, marbles, and even bottle tops, but as an adult, it has been the trading of ideas — learning about them and discussing them — that has sustained me. At the time of my postgraduate clinical training, it was accepted that we had a finite number of neurons, with largely fixed connections. There was little a brain-injured person could do for cognitive rehabilitation. It was understood that brain function improved spontaneously over the first six to twelve months, and that thereafter it plateaued, with only minor improvements in the coming years. It wasn't clear why this pattern occurred. The brain-injured person was a passenger in their rehabilitation, not the driver. The more diffuse the brain injury (strokes often cause diffuse injury), the more wide-ranging the damage and the resulting deficits, and the slower the recovery.

Doidge tells me the brain is plastic: new connections can be made between existing neurons, and sometimes new neurons can grow. Neuroplasticity happens by focusing the mind's attention in specific ways so that neurons that 'fire together, wire together'. Like a technician, we can change our brain's circuitry to suit our purposes. He mentions research by Alvaro Pascual-Leone, who, using the method of transcranial magnetic stimulation, demonstrated that blind people who learnt to read the raised dots in braille developed larger 'maps' in the motor cortex for the fingers used for 'reading' than for their non-reading fingers. The area devoted to these fingers in their motor cortex was also larger than the equivalent area in those who couldn't read braille. In the blind person, the occipital area of the brain, which is usually devoted to vision, gets taken over by other functions, such as sound.

I already have some understanding of brain neurology through my clinical training and my undergraduate major in zoology. But Doidge is telling me that I can apply this knowledge to myself. I'm surprised by how uplifted I feel as I absorb his ideas; it takes me out of my life's messiness. I want to know more.

Each neuron has a bulbous cell body — the grey matter of the brain — with dendrites, like spidery arms, sprouting off it. The rest of the neuron is made up of a long, thin fibre — the axon — that ends in numerous fist-shaped synapses. The dendrites and axon of a neuron can grow 'sprouts' to make new connections. A connection between neurons is most often made when the synapse of one comes into proximity to the dendrite of another. Information passes along the axon via an electrical impulse until it reaches the synapse. Here, neurotransmitters — the brain's chemical messengers — are released, and they move, in microseconds, through the minute gap between the synapse and the neuron it is cosying up to. As a newly made connection is reactivated over and over, the synapse and dendrite become sensitised to each other: they communicate more efficiently, like good friends.

A single neuron can connect to thousands of other neurons, and it is this capacity for neuronal connectedness that provides the landscape for neuroplasticity. I think of it as being akin to families. A small nuclear family has few members to draw on for knowledge, skills, and resources. If something happens to one or two of its members, the family unit is in jeopardy. The advantage for the nuclear family is that communication between them is quick. Yet an extended family, with multiple generations, draws on more resources. As long as they cooperate, this family has greater resilience and capability because of their wide-ranging connectedness.

The neurotransmitters in the brain — especially glutamate and gamma-aminobutyric acid (GABA) — act like neuronal 'on' and 'off' switches. Glutamate is excitatory, activating neurons, and GABA is inhibitory, reducing activity. This 'on' and 'off' process is the stepping-stone of neuroplastic change.

There are also many other neurotransmitters that modulate neuronal activity, enhancing neuroplasticity. Focusing attention on something activates the nucleus basalis (above the brain stem) to release acetylcholine, which diffuses throughout the brain, helping to sustain this attention. Dopamine is released when a reward is expected or a goal has been achieved. Noradrenaline is released when we come across something new, alerting our brain to take notice. Serotonin is active in creating the emotional quality of wellbeing. So the acts of focusing attention and experiencing novelty, reward, and emotional tone all help to cement neuroplastic changes.

When a number of neurons fire together, they become a neural network, threading its way throughout the brain. A particular memory is distributed through the brain as a living chemical and electrical trace, and the more neurons employed in the memory, the more secure it is. Recall of a memory reactivates the same neural network that responded to the original event; remembering requires the brain to pull together the sensory, motor, cognitive, and emotional components of the memory. The hippocampi, left and right, act as the managers in coordinating this recall.

The damage to my left hippocampus and temporal lobe probably explains my difficulty in recalling the names of things and people, and in remembering what others tell me. The recall of a complete memory is like a successfully baked cake. A cake requires flour, baking powder, eggs, butter, sugar, and flavouring, but if one of these ingredients is missing, the cake will still be cake-like, just not as complete. Similarly, when memory recall is altered by brain damage, there is still a memory, although it will be missing elements of the full memory.

The key thing I take from Doidge's book is that training — repeatedly reactivating a neural network — is necessary for long-lasting brain changes. This results in the brain learning a new skill, or new knowledge, which becomes more embedded with repetition.

Certain brain chemicals cement neuroplastic change by their actions on synapses and neurons. Brain-derived neurotropic factor (BDNF) is a protein that helps to maintain existing neurons and encourages the growth of new neurons and synapses; its supply in the brain is key to neuroplasticity. In the family analogy, BDNF is like the food the family members eat; the amount and quality of this food is critical to their wellbeing. Most neurons have a white sheath — myelin — surrounding the axon, the white matter of the brain. As the neuron becomes more active, the myelin sheath grows in thickness. This increases the speed at which an electrical impulse travels along the axon. BDNF facilitates myelinisation.

Driving a car requires focused attention on sensory, motor, and decision-making aspects. With practice, the neural networks involved in driving are more efficiently connected, thereby requiring less of the driver's conscious attention. Learning to drive well and safely is an activity with a high degree of focused attention and motivation, and both of these factors work to enhance the neuroplastic changes. I remember that as a learner driver, there were an overwhelming number of things to concentrate on. My hands gripped the steering wheel tightly — as if I was afraid that

I might somehow let go — and I could only look straight ahead. Conversation about anything other than driving was impossible. But in time, driving became automatic, and I could think and talk about anything, holding the steering wheel lightly, while keeping my eyes on the road.

Neuroplasticity not only means new connections between existing neurons. In a few parts of the brain, new neurons grow from stem cells, in a process called neurogenesis. This is the part that really excites me, because neurogenesis has been found to occur in the hippocampus. Can I restore my damaged left hippocampus and, therefore, get my memory working again?

I best understand the variations of neuroplasticity using a road analogy: new sections of road can be built to provide ways around blockages in the existing road network (synaptogenesis), completely new roads can be built (neurogenesis), and existing roads can be made easier to travel along through widening and resurfacing (myelinisation). The type of change governs the time required: synaptogenesis takes minutes to hours; neurogenesis takes weeks; and myelinisation takes months. Quick neuroplastic changes strengthen existing neural connections, while the slower, but longer-lasting, changes rely on the formation of new connections and new neurons.

My experience of how hard it is to maintain a conversation and how noxious certain sounds have become — some people's voices, music in cafes and shopping malls, and mechanical noises — points me to the belief that I have an auditory-processing problem. Doctor Small has told me that the damage to my brain, while most obvious in the occipital lobe, has also encroached into the temporal lobe on the left side, an area critical for understanding speech. This could explain why I forget what others tell me, and so my difficulty in keeping up in conversation: if I'm slow at taking in what people say, there's less chance of remembering what's been said. The words of new people, with their unfamiliar voices, requires more neural processing than do the familiar voices of family and friends.

I can hear speech all right, and I can make the sounds of the words in speech. But my brain finds it difficult to translate the sounds of speech into engrams — neural representations of words. And there's my difficulty with finding the right word, too. According to Doidge, this problem can be due to 'fuzzy engrams'.

It reminds me of my trip to Paris in my early twenties, when I relied on my schoolboy French. I could pick up a word or phrase here and there when listening to the native speakers, feeling as if I should be able to understand them, but I couldn't, really.

If my neural circuits for making others' speech intelligible were once dual-lane freeways, they are now single-lane highways; my brain is trying to handle the same amount of auditory traffic as before, but with reduced neural infrastructure. No wonder I'm drawn to quiet places, soft voices, and people who speak at a slower pace.

I begin to wonder: could I retrain my brain specifically to improve auditory processing? I have to find a way of getting my brain muscles working again — my and my family's future depends upon it. Doidge mentions Posit Science, a provider of training programs aimed at reducing cognitive decline in the elderly, and I decide to investigate further.

When I look at the Posit Science website, it says that they offer two programs designed to help the elderly with the common outcomes of cognitive decline associated with ageing. The InSight program improves visual processing. The Brain Fitness program improves auditory processing: forgetting names, slowness of thinking, difficulty in word retrieval, difficulty in deciphering speech, and fatigue in conversation. I'm not elderly, but this describes me perfectly. Could it work for me too?

RECOVERY

12

IT'S SEVEN MONTHS post-stroke: February 2010. I'm anxious to get on with my recovery, and the more I read, the more it seems like a computer-based cognitive training program is what I need.

I ask Doctor Small about such programs, but he knows nothing about them. Then I ask Doctor Mercer, the ophthalmologist, if computer-based training would improve my visual deficit, but she's not heard of this approach either. She recommends gentle exercise, such as walking along the beach, and tells me that I'll see improvement with the passage of time.

Doctor Small has recommended a neuropsychological assessment, so I decide to put off ordering a program until I speak with the neuropsychologist. A standard assessment includes a clinical interview and a series of tests that aim to measure a person's cognitive functioning, including short- and long-term memory (both verbal and visual), processing speed, spatial skills, conceptual thinking, decision-making ability, and learning capability. I am keen to know the extent of my cognitive capacities, like a high jumper wanting to know how high the bar can be set.

I call the neuropsychologist to let him know that I am a clinical psychologist and I have experience in neuropsychological testing. I'm aware that my prior knowledge of some of the tests might influence the results. He asks me what tests I've used in the past; he'll find alternatives, he says, and thanks me for making contact.

On the day of my appointment, I meet him in his city office, which is quiet and comfortable. The assessment, including the interview, takes almost three hours, with a break in the middle. In one test, he recites a list of eight pairs of unrelated words and asks me to repeat them back to him. After the first trial, I correctly remember two of the eight word pairs. He repeats the list three more times, and after each trial, I repeat back the word pairs I can remember. By the end of the fourth trial, I have only learnt one extra pair — three all up. I'm trying really hard, but the words just don't stick in my mind. This is the type of test I've given many times before, and I know that mine is a poor result: I should've learnt most of the pairs by now.

The neuropsychologist reads out two short stories. Then he asks questions about the content of the stories, and, some time later, asks me to recall as much as I can. Once again, my performance is poor, even though I think I've administered one of the stories myself in testing clients.

My memory for faces and my working memory (working memory typically lasts twenty to thirty seconds) are poor. My visual memory is good, and my vocabulary and speed of processing on visual tracking tests is in the superior range.

My uneven results across the different tests points to an organic cause for my cognitive deficits that's consistent with a brain injury. The neuropsychologist says that the high level of chronic stress and my post-traumatic stress disorder could also have caused the cognitive deterioration. He tells me that he went to a professional-development seminar where the presenter showed the test results of four cases: two with post-traumatic stress and two with brain injury. The participants in the seminar were unable to distinguish the non-brain-injured from the brain-injured by the test results alone.

In his opinion, there's no way I could return to clinical work: my poor memory, difficulties with auditory processing, and mental fatigue would preclude it. Also, my past exposure to trauma puts me at risk of being re-triggered.

I ask the neuropsychologist about computer-based cognitive training. His only knowledge is of the training games available for the Nintendo DS, a small handheld gaming gadget I've seen children using.

On my way home I look at these brain games in a store, and they're simplistic compared with the sophistication of the Posit Science programs. I reach a conclusion: I'm going to have to design my own rehabilitation program. And the first step in that will be brain training.

But what program do I choose?

An internet search reveals there are few commercially available programs, and what little scientific research I can find on computer-based cognitive training points to Posit Science having the most backing for its claims. I locate the Australian distributor, Alzheimer's Australia, and call them. Their adviser, Matthew, tells me that they decided on Posit Science's programs based upon an exhaustive review of existing cognitive-training programs they undertook in 2008.

I will only manage one of Posit Science's two programs with my limited mental energy. I've lost a quarter of my visual field, and the InSight program, which trains visual processing, could help with that. But what hampers me most in daily life? I'm driving okay, and my visual memory is good. All in all, my visual deficit does not restrict me greatly. By comparison, I think of how tiring it is to understand others' speech, and how I need to think out each word before I say it, as though English is now a second language.

Matthew says that studies with older people have shown that those using the Brain Fitness program double their processing speed and gain more than ten years on their scores in standardised measures of memory and attention, leading to improved comprehension. Yes, this is what I need most. According to Matthew, I'm not the usual type of person who enquires about the programs, but he can't think of a reason why I shouldn't give it a go.

I pay for the two-person version of Brain Fitness, thinking that Anna might like to do it too.

BEFORE I CAN get started on the Brain Fitness program, there is something else I need to deal with. In late March, I walk out of a lift and taste the dry, air-conditioned air. I'm in a shining chamber of marble — almost too shiny to look at.

I haven't been looking forward to today.

Over to my left, a receptionist with perfect hair and a pinched look of concentration sits behind a highly polished counter. She wears a telephone headset: the type that looks like a hairband, where the incoming call goes into only one ear, leaving the other ear free to hear external sounds. It is to this ear that I direct my enquiry. 'I'm here for a meeting with Mr Tsanov?'

'He's in conference,' she says. Gesturing behind me with a flick of her head, she adds, 'Take a seat.'

In the seating area, there is a long coffee table with an etched-glass surface. Set upon this are two tall, fluted vases holding long-stemmed gladioli with pink and red blooms: one at each end of the table. Placed precisely in the centre is a stainless-steel tray, upon which sits a plump glass jug, filled with water to its throat, where blocks of ice congregate. The ice and the beads of condensation suggest that the water is nicely chilled, untouched by anyone this morning. Circling the jug, like a small band of kindergarten children — upright and attentive, as if waiting for their teacher's instructions — are eight clean glasses. I have a dry mouth after the walk from the bus terminal, but I don't want to disturb this perfect arrangement, perhaps put together by the woman with the perfect hair. And I don't want to drink *their* water and feel myself coming under *their* sway. Instead I sit on one of the heavy leather sofas with fleshy armrests, the two facing off like bull-mastiffs.

I reflect on what's happened over the past few months. A property we'd purchased off the plan, which was going to be a retirement apartment for Anna's parents, was due for settlement late last year, but of course we couldn't pay for it. A property investor had recommended Donald Trump's book *Trump Never Give Up*. When Trump faced bankruptcy, he acted on faith that he would work his way through it. He stressed that it was important not to ever give up. I followed Trump's advice to make personal contact with creditors: I phoned the developer and told him, before settlement was due, of our financial position. He sounded accommodating, saying, 'Get your lawyer to write to us and state your case.' I wasn't keen on this approach: lawyers meant that negotiations would become adversarial and expensive. But we couldn't dictate the process, so what else could we do? Doom and Gloom wrote to the developer with supporting financial and medical documentation, and all we got for our trouble was a legal letter of demand from the developer's lawyers — settle the purchase or we'd be up for damages. I tried making contact with the developer again, but never got a return call.

Then, a few weeks ago, the developer called us out of the blue. He apologised, saying that he hadn't intended things to get adversarial: his hands had been tied by his overseas bank after he had defaulted on his development loan. But now he had the bank's authority to offer us a substantial reduction in price. Many purchasers were defaulting, and the dreadful market conditions were making it impossible for them to get finance. I told him that we might've been able to accept his offer when I spoke with him last year, but our cash reserves had since been swallowed by legal fees, loan repayments, and living costs. He said he couldn't help us any further, and, giving his lawyers a blast, complained that the fees they were charging him were outrageous. He added that he was madly trying to sell his own properties to maintain the payments on his development loan.

We were at a stalemate. I could only conclude that we would be sued by the developer for damages.

I am here, in the offices of this major law firm, because I am fighting over money, which Anna and I will need even more acutely if the developer does sue us. I had asked my income-protection insurer to backdate my claim from the date I had actually stopped work: two years before I lodged the claim. With legal costs, and large ongoing loan repayments to make each month, Anna and I needed more funds. But the insurer didn't accept my request, saying that I had not sought the advice of a medical practitioner in my treatment prior to making the claim.

This was true, in a sense. I had 'referred' myself to a clinical psychologist, Wayne. If I had asked my GP to write a referral letter for Wayne, I would have been following the advice of a medical practitioner, but I hadn't thought it necessary to do so. Although I had sought the advice of my former colleague, Ian, the insurer saw this as an 'informal arrangement' that did not meet their definition of 'under the direction of a medical practitioner'.

To force the insurer's hand, I sought the assistance of two lawyers, Simon and Andrew. They were confident of achieving a positive outcome, and we filed a statement of claim in the

Supreme Court. We are here today, nine months after filing, at the invitation of the insurer's lawyer, to try to reach a negotiated settlement. If we don't settle today, we will do battle at a court hearing in two months' time, but Anna and I will not hold out financially for another two months.

My lawyers arrive. Simon is the solicitor. He has a hale and hearty manner, ginger hair, and a stocky build. Then there's Andrew: tall and lean, with a conservative coiffure and the intellectual manner of a barrister. He makes an effort to be friendly, in a self-conscious way. He's dressed in an impeccable black suit and emits measured assurance.

Andrew gives me a rundown of what he's expecting for the morning, and he sounds confident. He's like a sports coach giving final instructions before the game. 'I'll do the talking. You don't need to say anything. If it gets too uncomfortable during the conference, you can excuse yourself and leave the room.'

'I'd like to be present,' I tell him.

He thinks that today will be procedural, and we'll have a settlement by the day's end. 'We're here to support you. We're on your side,' he reassures me.

Off either side of the waiting area are two conference rooms. Andrew talks to the woman with the perfect hair, comes back, and directs us into the conference room to the receptionist's right. There is a jug of chilled water here too — this time set unceremoniously on the kitchenette sink — and now I am too thirsty for mental games of defiance, so I help myself to a glass. As we stand by the kitchenette, marking time, I can see through the open door into the reception area. I hear the sound of the lift doors open, and out walks the man I suspect is my support person, Craig, although I haven't met him yet — we've only spoken over the phone.

Craig has worked in the life-insurance business and knows my policy inside out. I wanted him to come today and explain, in non-legal terms, what is happening and to help me with the decisions I will need to make. Since the stroke, complex decision-making has become terrifying; and, although I have no particular reason to distrust Simon and Andrew, over the last year I've developed a dislike for lawyers. Craig is dressed in a suit, and has a few curious features: a small stud in his left ear that shines like a diamond, rakish hair, and a pink tie that doesn't coordinate with anything else he is wearing. The other noticeable feature is his physicality: he's big, with the chest of a bull. I'd say he is in his fifties, both fit and strong.

Andrew has been looking at his watch off and on, and, once Craig has introduced himself, says, 'What's going on? We were meant to start the meeting at ten o'clock. It's now ten twenty-five. They've had weeks to prepare for today. What are they playing at?'

We wait another five minutes or so, and then Simon gets a call. I think it's about the case, so my ears prick up when his voice gets noticeably louder. But instead I hear him say, 'Well, when I put the washing on the line this morning, it wasn't raining. I can't predict when it's going to rain. What do you expect me to do about it? I'm in Sydney … Look, I'll see you tonight.' He snaps his phone shut and says to none of us in particular, 'She expects me to be a weather forecaster!'

Finally, there is movement from the conference room opposite. The door opens and out comes a thin man with a beaked nose and heavy, square glasses. He says he is Mr Tsanov, the barrister engaged by the insurer. Simon introduces each of us in turn, and we shake hands. Tsanov gives me a second's worth of eye-gaze — no smile — and a short hello in a plummy, resonant voice. Following in his wake is a woman who almost curtsies in deference behind him, and who is introduced as the insurer's in-house lawyer. Her expression when introduced to me is sympathetic; I wish we were dealing directly with her instead.

All six of us sit at the long conference table. I am at the head, with Tsanov to my left, and his assistant lawyer in the next chair along. Andrew is off to my right, with Simon the furthest away. My large support person sits in the wedge of space between Andrew and me. Tsanov has insisted that if Craig is to be present during the conference, he is not allowed to speak. We have no choice but to accept this condition.

The barristers square off with each other, and although it is an informal conference, I can see them mentally putting on their wigs and gowns, adjusting themselves to sit taller in their chairs. Tsanov searches silently, unhurriedly — all eyes on him — through a folder of documents resting on the table. Then he leans forward, as if seeking intimacy (his tie bent by the edge of the table, eye contact only with Andrew), and, without any preliminaries, asks, 'What is your client's position … what is he seeking?'

Henceforth, I become invisible.

Andrew refers to a sheet of paper, naming each item in my claim, like an old-fashioned greengrocer with a list on a notepad, written with a pencil grabbed from behind his ear. Unpaid monthly benefits from this date to that date, interest, costs, refund of paid premiums, and interest on interest. He cites the dollar amount of each item, and finally, looking to Mr Tsanov as if expecting immediate payment, announces the grand total. Tsanov gives this summation a moment's disdainful consideration, bows his head to look at his documents again, and the real tussle begins.

I have already decided that I will tune out during most of the conference and maintain a look of equanimity. I think I know how this game is played, based on stories from past clients and from television shows: the other side's lawyer tries to agitate you, needle you, and catch you out in some way.

'Your client's claim has a number of flaws, I'm afraid,' Mr Tsanov begins. His tone sounds reasonable, even considerate, but the import of what he is saying has a growing malevolence about it.

'To suggest that Doctor Somerville, a clinical psychologist, is a medical doctor is absurd. I've looked at the registration requirements for medical practitioners in this state …' he intones. He's referring to Wayne. I reassure myself that Andrew will deal with all this. As Tsanov continues, I notice the thinness of his long neck, with its protruding Adam's apple that moves skittishly as he speaks. Along with his small head and the thick glasses that magnify his eyes, I can't help but imagine a turkey. As I mentally withdraw from attending to the meaning of his words, his speech begins to sound like the *gobble, gobble* of a turkey.

I build a life story for Turkey Neck. I imagine that he attended a private boys school, spending his lunchtimes in the library looking up reference or special-interest books. He was probably a member of the chess club, and no doubt excelled on the debating team. Now I'm receiving the brunt of his debating skills. With his slight frame, average height, and glasses, I imagine he avoided the parts of the school playground where the sporty boys hung out. I envision that basketballs, thrown 'accidentally', would sometimes hit him on the side of the head, knocking off and breaking his glasses, and they would need to be patched up with tape until he made it home and got his spare pair, his parents resigned to ongoing optometry bills.

Now, Turkey Neck is looking and sounding like an old-time headmaster chastising his pupils: so confident, so superior. He is taking my legal team to task. 'Of course, the logic of your argument is ridiculous. Doctor Roland has clearly managed his own affairs and has not relied on a medical practitioner's directions at all. He didn't even mention his condition to his GP for at least a year and a half, and yet consulted him on a number of occasions for other ailments.' As he says this, I see him, out of the corner of my eye, glance towards me, as though to determine if his remark has provoked a reaction from me. I remain impassive.

'I don't want to insult Doctor Roland's intelligence,' he says, and then goes on to detail — in that tone of reasonableness — how unintelligent I have been.

What's going on? This is supposed to be a negotiation. Andrew had told me that we were invited to a *settlement* conference; he didn't say it was going to be adversarial. But Turkey Neck is having a good old poke at my claim, and is barbecuing my legal team in the process. He rummages around in our box of arguments, picks up each one, and holds it at arm's length with pinched fingers, as if saying, *You mean this is an argument, this fragment?* and then drops it back into the box. I feel like squeezing his turkey neck.

My team looks off-colour. Andrew spits out short retaliations now and then, and occasionally rises to launch a salvo. But I can see that he is rattled, and his retorts are smothered by Turkey Neck's words. He is leaning back in his chair, like someone facing a barking dog. Simon, although not directly in the line of Turkey Neck's assault, exhibits similar body language.

Turkey Neck refers to a letter that apparently says something that disadvantages my case. Simon says, 'Oh, I'm not sure I've seen that.' He flaps through his folder of documents, the size of the city telephone directory, as if somehow the offending page will float out — magically, as in *Harry Potter* — into his hands. After a few minutes of flapping, he says, 'I can't seem to find it. I'm not sure that we've received that.'

At this, Turkey Neck's expression seems to say, *I thought as much … total incompetence.*

As my confidence in my legal team wanes, I let go of the idea of strangling Turkey Neck — it's stirring up a disconcerting feeling of anguish. I tune out again. As far as Turkey Neck is aware,

I remain unmoved, staring out the window at the end of the room. Fortunately, I can see a large patch of perfect blue sky. In my mind, it becomes the blue of the ocean. I am swimming in the bay near home. I can see turtles and fish, and feel the sand, gritty in my bathers, as I stand up in the surf after being tossed around by breakers while coming into shore. My friends are there, the sun is out, and I feel all right.

I'm brought back when I hear the barristers declaring that they'll take a break and have a private conference because 'Doctor Roland is probably tired out'. I do welcome this, and the barristers and solicitors trundle off to the other conference room while I'm left alone with Craig.

Craig says that he is not in favour of adversarial tactics: it puts the insurer offside. Taking the insurer to court, as I have, makes them less willing to negotiate. But I did not have his advice before. He acknowledges that there's no point in dwelling on it. He suggests a compromise, which sounds reasonable to me. I agree, and make a note to discuss it with Simon and Andrew.

But before they come back, I change the topic of conversation. I'm sick of talking about legal complexities, and I'd like to get to know a bit about Craig. Somehow, he gets onto telling me that he is a Remote Area Firefighting Team volunteer. I haven't heard of this before. 'It's a specialist unit with the Rural Fire Service. We go where normal fire vehicles can't get in. We're dropped in by helicopter.' He explains how they set up new fire fronts to fight the existing one.

'What sort of training do you need for this?' I ask.

He says that because they fly in helicopters, they need to be prepared to land in water. The fire service has a metal compartment the size of a helicopter cabin, and they drop this into the swimming pool used at the Sydney Olympics. When a helicopter lands in water, it can turn upside down, so to prepare for this, the volunteers are strapped into the compartment before it is dropped into the water upside down. They don't have oxygen tanks, so they rely on holding their breath while they extricate themselves from the compartment and swim to the surface.

'This isn't easy,' he says, 'and it plays with your mind. Some can't handle it. The key thing the instructor told us was that we have to know how to exit the compartment *before* we land in the water. We have to have our hand or arm on the exit lever *before* we crash, because under the water — in most conditions — you can't see in front of your face.'

As he is describing this, I'm aware of a growing sense of discomfort. I'm being transported into my nightmare, where the family and I are driving in the failing daylight, and the van runs off a bridge and into a river. What he has just told me confirms that I would not be able to see the children in the back seat. I could not even know exactly where they were. I would have enough trouble trying to get out myself, without trying to help them. In my dream I could see them, and I would struggle to free them from their seatbelts, pushing them out the door to the surface. But with what Craig has just told me, they would surely drown.

I thought I had dealt with this nightmare in therapy, put it behind me. But here it is, stalking me again.

I'm shaking, and I don't know if Craig notices. I have to get out of the room. I interrupt him and say that I have to go out for air. There's no one in the waiting room. I grab a glass of water and stand by the window. *We are sinking into the river, the water is pouring in through the windows and the gaps of the car, and now I can't see the kids.* This is bad. My chest is heaving. My breath is coming in waves. I try to jemmy the drowning images out of my head. I breathe and breathe and breathe, willing my breath to slow down; I press my face against the window, drinking the water slowly and hoping the chill of it will bring me back to the present. *It's not really happening. You're in a lawyer's office. You're all right. It's daytime.*

After a while, the images and the physical sensations of the nightmare loosen their hold. It's like being on a train that is slowing down as it comes into the station, and the snapshot view of the platform through the window lasts longer and longer. The view of the skyscrapers through the glass begins to last longer, becoming more solid, and finally fixed. At last, I really do feel I am in the waiting room of a high-rise building, in the office of a legal firm, with a turkey-necked barrister who wants to squash me. And the thought of returning to do battle with him isn't as gruesome as the nightmare.

After my legal team has knocked heads in 'secret lawyers' business' with Turkey Neck and his sidekick, they emerge from the second conference room. Andrew declares, 'It's useless. We didn't come here to debate the case, but to settle. We've wasted our time. We may as well all go home.'

All this trouble for nothing.

TO MY RELIEF, the calls between Simon and Turkey Neck start the next business day. Turkey Neck is finally in the mood to settle. Simon rings me almost every day, telling me with excitement that Turkey Neck has raised his offer, and we discuss a counteroffer. I imagine Simon and Turkey Neck as two medieval knights jousting, their telephone receivers their lances.

I wonder why we have to play this silly game. Why doesn't the insurer just say what dollar amount they are prepared to go up to, and we can either accept it or not? In the end, it will cost the insurer much more to play this game than if they had been reasonable with me in the beginning, before the lawyers got involved; and I will get less out of the settlement because of the legal fees I'll have to pay. The only winners are the lawyers.

Some days later, we get an offer. It is at the bottom end of the range of what Simon and Andrew thought we could achieve. I have to accept it. It is two-thirds of what we were claiming, and when I get my lawyer's bill, I'm left with two-thirds of this amount again. The money will allow Anna and I to meet all our loan repayments and expenses for the next six months or so. After this, the battle for financial survival will start again.

AT THE START of April, some better news arrives — in the form of my Brain Fitness program, which turns up in the mail. I'm eager to get started. I need to get my brain working again; I'm not going to get out of this mess without it. Anna says she has no time to do the program, so I'm on my own.

For optimum results, Posit Science recommends the completion of forty hours of training over eight weeks — five days a week, one hour per session. On the first day, I only manage thirty minutes; after this, rubber brain threatens to overtake me. The next day is the same, as is the following. This shows me that I'm working to my limit. At this rate, it's going to take me four months or more to complete it.

The program concentrates on building the basic auditory skills first (pitch and phonemes), and then the components of speech (syllables and sentences), and finally comprehension (narratives). It contains six different exercises, which I work through progressively. The first, 'High or Low', trains for pitch in speech, drawing on the frequencies found in spoken consonants and vowels. It does this by frequency sweeps: a computer-generated sound begins low and rises in pitch, or begins high and lowers. It's a sound like a zipper being opened or closed quickly. I listen to the pairs of sweeps in my headphones. I have to decide if each sweep in a pair has gone upward or downward, and use my mouse to click on the correct sequence of up and down arrows on the screen. The program picks up on my progress, making the sweeps quicker and reducing the time gap between them. The brain needs to be pushed beyond its current limit to improve. I quickly reach my threshold, and it becomes difficult to tell whether the sweep is going up or down.

The second exercise, 'Tell Us Apart', uses phonemes — the individual sounds that make up words. In the word 'dog', for example, there are *d*, *o*, and *g* sounds. The program presents similar-sounding phonemes: for example, the sounds 'dah' and 'gah'. The sounds are hard to tell apart. I perform very badly.

The program notes indicate that the voice saying the phonemes has been modified to alter the speed at which the phonemes are said and how much emphasis is put on the consonant. A faster speed and less emphasis makes it harder to tell them apart. Some phonemes are easier for me to work out than others. For some reason, my brain finds certain consonants harder to process than others. This shows me why I find the comprehension of others' speech, such as that of the lawyers, so tiring: my brain is working overtime, trying to make sense out of fuzzy engrams.

There is one exercise I enjoy. 'Match It' is like the card game Memory. A matrix of cards is presented facedown on the screen. Each card has a syllable associated with it, and within the matrix there are pairs of syllables. Some of the syllables are dissimilar in sound: for example, 'baa', 'fo', and 'pu'. Other syllables sound similar: for example, 'sho', 'stu', and 'sa'. I am allowed to click on two cards, one after the other, and hear the sounds they represent being spoken. I work my way through the matrix, activating each card to find the pairs. This is training my working memory for spoken words, with a spatial-memory component. The matrices increase from eight to sixteen to twenty-four to thirty cards. I excel at 'Match It', compared with the other exercises; it's encouraging to still be good at something. Perhaps it shows that when I can use my visual memory, it aids my overall memory.

As I progress onto the larger matrices, I notice that I can let go of mentally rehearsing the sequence of sounds I've just heard. Instead, when I click on a new card to hear a sound I've heard before, I let my mouse hand drift over to the card that 'feels' like the match and click on this. Most often, it is correct. Somehow, my subconscious processing has become faster and more accurate.

I'd love to spend more time doing the 'Match It' exercise, but the Brain Fitness program, like a good teacher, soon learns my weak areas and focuses on the exercises that most challenge my brain. One of these is 'Sound Replay'. It presents syllables such as 'baa', 'fo', and 'laa' as a memory-span exercise, asking me to remember a series of such sounds, as if I am learning a list. The voice names a random sequence of syllables, starting with two and then moving on to three, four, five, and more. I need to indicate which syllables were said, and in what order. This is training my capacity to discriminate sounds and is building my auditory working memory. It reminds me of the paired-words test I did with the neuropsychologist. Like then, I do poorly on this exercise — the sounds quickly enter the fog. I can only remember two syllables, and occasionally three syllables, for a long time.

'Sound Replay' and 'Tell Us Apart' are the hardest of all the exercises, for me. After the first few weeks of training, I remain on the lower levels for these exercises, wondering how I'm ever going to progress beyond this.

In 'Listen and Do', I am presented with the visuals of a street scene that contains people, animals, and objects in the foreground, and buildings in the background. I hear a set of instructions: a sequence of people and objects that I am to click on. Once the instructions are given, I need to click on the objects in the same order. As the exercise advances to higher levels, I have to move a person or an animal to a new location. (For example, 'Move the redheaded girl to the left of the brunette girl', or 'Move the black dog to the right of the hospital.')

I'm okay with this exercise, once I develop a strategy for it. I draw imaginary lines between each named object, giving me a visual shape I can remember, which aids in recall. This resembles my real-life task of visualising a mental list of things to do and staying on track until they are all completed.

The final exercise, 'Storyteller', is the most enjoyable. The voice tells a story of everyday interactions and events happening between people. At its completion, I need to answer ten, fifteen, or twenty questions about the story I've just heard. The answers are in multiple-choice form. The stories, five in all, become progressively more complex. This is training short-term memory and comprehension — being able to remember and understand details in spoken conversation. I recall the short stories the neuropsychologist read to me and how poor my memory was; I can see already that I've definitely improved on this type of task since then.

Progress bars appear on the screen during each exercise, and I strain to get to the next level of difficulty. There's a sense of achievement (a dopamine hit) when I reach a new level, and the program rewards me with animated fireworks and music. At the end of every session I can access a summary page, which lets me know how I'm going with each exercise. This is reassuring (even though it shows how poorly I'm performing in most exercises): it gives me a baseline from which I can see increments of improvement.

As I get into the rhythm of the program, most days I advance on an exercise or stay at the same level. But sometimes I have an off day. That's when I see, in the form of the progress bar, how dramatically my performance drops off when I'm mentally or physically fatigued — the clearest indication so far that fatigue really does affect my day-to-day capacity to function.

After one month of doing thirty minutes most days of the week, I have progressed up the ladder to some degree in all exercises. But I haven't noticed a great difference in my auditory processing outside these exercises. I'm a little discouraged by this. Yet the cheerful male voice that explains what each exercise does and why it is useful shows a picture of the brain with coloured lines connecting the areas that each exercise works on, reassuring me that I am making new neural connections. If it doesn't work, I've only wasted some time and some dollars.

I'm consistent with the Brain Fitness training, even when I'm feeling off, or stressed and anxious. At first I do the session mid-afternoon, before I get rubber brain for the day, and after I've done a decent amount of wrangling with lawyers, real-estate agents, and creditors. But after a while, I realise that I make better progress if it is done earlier in the day. I change tack, doing the training session just before lunch. Yet on the mornings when I attend a medical or psychological appointment, I can't do this: these sessions knock me around for most of the day.

By the six-week mark, I've noticed a real difference. My progress has been gradual, but all of a sudden the world is easier to comprehend, as if a door has opened. Other people's speech seems clearer; it's less of a strain to listen to them, and easier to understand what's being said. Everyday social conversation with another person is simpler, and I'm less fatigued afterwards. I still lose track in longer, more involved conversations, and I get fatigued after group conversations, but I'm confident now that I'm showing real improvement. My brain is coming back online and is starting to work with me.

In time, I also notice that the individual notes in the music that Nick and I play, or the sounds that other musicians make, stand out more than they have since my stroke. Phone calls remain taxing, and I'm still easily distracted by noise or other disruptions. But I'm remembering the names of familiar acquaintances easily now, and I think that my sense of direction has improved a little too — although it's nothing like it used to be.

I catch up with Wayne again and, after we've talked for a while, he says, 'You're looking brighter. Do you realise you haven't asked me once to repeat a question?' Gee, is the difference that obvious? I can't help but smile.

13

I BEGIN TO wonder if I could be doing more for my brain rehabilitation; it doesn't feel as if I'm doing a lot, except for the Brain Fitness program.

I call Chris, a regional case manager with the Brain Injury Rehabilitation Service. In the past, she's referred brain-injured clients to me for psychotherapy. Her service can't take me on as a formal client because they are only funded for traumatic brain-injury cases, not for strokes; there is no service in our region for stroke victims who can 'walk and talk'. But, she says, she'd love to see me.

When I arrive at the old house that is their headquarters, Chris introduces me to another case manager: Leanne, a speech pathologist. We sit around a table in a heavily furnished room that serves as their meeting room. I explain what's happened to me and pass over a copy of the neuropsychologist's report. I try to summarise his findings, emphasising my poor auditory processing and verbal memory.

Chris notices the neuropsychologist's comment that I have

mild cognitive impairment. 'Mild cognitive impairment is clinically significant.'

I want to know what chance I have of recovery. 'Most improvement will happen in the first two years,' says Leanne, and Chris adds, 'But some are still recovering up to ten years later.'

I mention that I'm engaging in psychotherapy and meditation, and doing the Brain Fitness program. 'Is there anything else I could be trying?' I ask.

They each ask questions about what I'm doing day to day.

For a while I keep up with their conversation. But soon, with the two of them coming at me, their words become an auditory blur. 'Would you like us to speak more slowly?' Leanne asks, looking at me intently. 'You seem overwhelmed.'

Yes, I would. My brain is like an underpowered engine trying to make it up a steep hill.

They slow down, but without a hint of annoyance. Am I keeping a diary with daily tasks and appointments, and crossing them off immediately after they're done, so I know where I'm up to if I get thrown off track? Am I putting my keys and wallet in a bowl in one place? Do I call regarding upcoming medical appointments, to confirm when and where they will be? Am I making physical changes in my environment to make things easier and to set up reminders? Do I carry earplugs to protect against noise? Am I exercising? Chris says, 'The main issue for the brain-injured is getting overloaded with information. In time, you'll work out your own strategies.'

'I've noticed you searching for words as we've been talking, using four or five sentences to say what could be said in one,' says Leanne. 'Finding roundabout ways to get where you want to go. Circumlocution.'

'Yes,' I agree, relieved to hear her name this. 'Like, I might want to say, "I've left my wallet at home," but I mightn't be able to think of the word "home". Instead, I'll say, "I've left my wallet where I live."' Often I'll keep talking, hoping that I'll get to the point eventually. When I've mentioned this difficulty to others, they say that I sound fine to them. I'm realising that part of my 'problem' is that I look and sound normal to others; my deficits are invisible.

'I think it would be worth doing a speech assessment,' says Leanne. She will arrange for a private speech pathologist to do this and recommend to my insurer that they pay for it, referring to the neuropsychologist's report.

Other than this, they both conclude, I'm doing a lot of the right things: I'm keeping a diary and to-do lists; I'm exercising. They haven't heard of the Brain Fitness program, but encourage me to keep it up.

'I'd like to be useful again,' I say. 'Maybe do some voluntary work.'

'You could assist a researcher,' says Chris, 'where you can work at your own pace in your own time. You have a research background.' She gives me a booklet from a conference on brain impairment she's attended recently. All the speakers are listed, along with their contact details. I could follow up with some of these people, she suggests, and highlights a few of them. She'd like the booklet returned when I'm done.

Leanne mentions the university health-research unit nearby as another option. 'They do a lot of dynamic research in mental health,' she says. As we wrap up, she hands me a piece of paper and I see that she has summarised our discussion, with the recommendations they have made, in point form. It's a thoughtful and very useful gesture.

I get up to leave and Chris grabs a book: *Over My Head* by Claudia L. Osborn — a memoir by a medical doctor who suffered a brain injury after being hit by a car while riding her bike. 'You might find this interesting,' she says. 'And call me any time. I'd like to know how you're going.'

When I look at my watch outside, it has been an hour and a half since I arrived. I've got rubber brain, but I'm pleased: I'm doing all I can, and I've found people who understand and *want* to help. I didn't feel abnormal with them.

I'd planned to do some supermarket shopping after the meeting, so I drive to a nearby plaza I've been to before. Parking underneath, I go up the escalator to the shops. I bring Chris's booklet with me; I'll stop in a cafe and go through it.

When I enter the plaza, the noise and crowds are an assault on my brain. I was being ambitious thinking I could look through the booklet here. I go straight into the supermarket. I've forgotten to bring the shopping list, of course, so I pick up what I think we need.

After the shop, I push my trolley into the parking area and realise that I have no idea of where I left the car. It's only one level of parking, but it's a large area. I see now that the thick columns holding up the ceiling of this concrete cavern are painted different colours for each section: lime green, rust red, mauve, sea blue. I should have made a note of which colour was on the column closest to me. I have no recognisable landmark on which to call. All I can think to do is to start on one side of the huge parking lot and zigzag up and down the rows, hoping that I'll chance upon my car.

As I move down the rows, the sounds — slamming doors, rattling trolley wheels, loud voices, the *vroom* of car engines — pierce my brain like knives. Now I really am suffering sore brain. I hear a loud, mechanical hum above me; at first I think it's from the long pipes, painted grey, criss-crossing between fluorescent tubes of light, and then I realise it's the invisible air-conditioning system. The petrol fumes are fire in my nostrils. I'm getting overwhelmed, but there's no quiet spot to stop and clear my head.

I keep pushing. I'm in the pink section now, but I'm not sure if I've passed this way already.

Soon it's been about twenty minutes and I'm close to breaking down. Am I going to be lost in here forever? Should I ask one of the people scurrying towards the escalator or to their cars for help? But what do I say — 'Have you seen my car?' I'm not even sure I could pull out the words to describe it. I'm going to end up standing by the trolley blubbering, like a lost child. I've forgotten my mobile, so I can't even call home. How will Anna know where I am when I don't return?

Suddenly, there it is: the familiar green roof of the car. I rush over and quickly unpack the shopping, shove the trolley into the nearest bay, and get into the front seat, closing my eyes. Just until I feel human again.

The booklet! The one Chris wanted returned.

I don't remember taking it out of the trolley.

Reluctantly, I leave the car to go and search the trolleys lined up in the bay. Nothing but other people's shopping dockets. I walk towards the escalators, remembering to look back several times to memorise the view I will see upon my return. As I go up, a child coming down the opposite escalator starts squalling — the beginnings of a tantrum. Daggers pierce my brain. I walk towards the supermarket, the uniform white tiles under my feet indifferent to my plight, gleaming with perfection.

I have to hurry or someone will pick it up. I go to the fruit and vegetable bays. I might have put the booklet down while filling plastic bags. The ordered arrangements of shape and colour jar with the chaos in my mind, but I need to find that booklet.

Nothing.

I go to the service desk and wait among the checkouts, which are clicking and beeping, the bored customers daydreaming, while my brain is screaming, *Get me out of here!* Has someone handed in a booklet? No one has, so I leave my telephone number and turn away. I'm in a shopping mall, for God's sake, but it's become a warzone.

I feel awful about losing Chris's booklet — about how incompetent I have become and how hard it is to conduct a simple trip to the supermarket. I am at the mercy of my capricious brain.

I drive home with the tinned tomatoes and the kidney beans.

A FEW DAYS after the supermarket trip, I bump into a friend in the street. He asks me how I'm recovering. I dread this question: it's so general. I don't know where to start or how much detail to give. I don't remember what I've told him before, or if what I think I've told him is what I've in fact told someone else.

I'm getting better, I say, but I lose things, like my car keys and wallet, and I get lost in car parks.

'I think we all notice these changes as we get older,' he says.

'No, it's different from that,' I say.

How can I begin to explain what happened at the shopping plaza? I don't have the mental energy. I know that the week before I had the stroke these things didn't happen, and the week after, they did. It's that stark.

It's not from a lack of caring, this type of comment, but some people don't get it; they can't imagine their brain suddenly not being there in the way it used to be. It would be outrageous to say to someone who has a broken leg and is having trouble keeping up that this is something we all experience as we age, or that they could fix their impediment if they really wanted to. I want people to simply accept it when I say, 'I can't do this like I used to,' or 'This is too hard for me now, even though it looks simple,' without trying to downgrade my experience into something they can fathom.

The next morning, I pick up *Over My Head*, the book that Chris gave me. It's a wonderful balm for what's just happened with my friend. The author moved to New York to begin a cognitive rehabilitation program. In the prologue, she describes the disaster of her first morning: getting herself ready to go out, giving her bus fare away to a beggar, and riding public transport while her memory and concentration are shot. Yes, that's me, but she's even worse.

Over the next week, I lap up the pages like a thirsty dog. It describes the misery she went through, the bewilderment she — and the others she went to rehabilitation with — experienced as she found out that things weren't the way they used to be. She writes of 'flooding': becoming overwhelmed and unable to make sense of what is going on, which sounds like my sense of 'rubber brain' and 'fog brain' combined. Claudia L. Osborn understands what I'm going through. I wish everyone could read this book; it explains so much. It puts into words — the words that I now don't have easy access to — my own experience.

I decide that most people (including health professionals, except for those who work with the brain-injured) are never going to understand, and I should give up expecting them to — it's hard for them to comprehend a deficit that they can't see or imagine.

I have already found that I gravitate towards others who have had their brain give up on them in some way. They can tune in to what I'm experiencing without any need for me to explain.

Like Doug at my ocean-swimming sessions.

Doug is a retired builder in his sixties who was conscripted into the Australian army during the Vietnam War. After he was discharged, he ran a high-pressure commercial shop-fitting and construction business in Sydney before it became too much. He had a 'nervous breakdown', as he called it, and subsequently left his wife. 'Best thing for both of us,' he told me.

Doug moved to the coast and bought a house. For a year he only slept, went for walks along the beach, and had regular sessions with an acupuncturist who, he said, was also a good listener. He's been getting out more frequently for a while now, and he often swims with the Stingrays. 'It gives me that social outlet. Gets me out of myself.'

I like to chat with Doug on the surf-club deck after the ocean swim. We share updates on our 'mental breakdowns' as easily as talking about the weather.

'How are you, Doug?' I asked him in one of our early conversations; he was bending over a backpack by his feet. He pulled out a comb, flicked me a glance as he straightened, and turned to look out over the sea while he slicked back his hair: Doug didn't rush.

Finally he turned to me and said, 'Dave, I've been up and down like a honeymooner's prick lately. I think I need to get away. Life's a shit sometimes, isn't it?' He put away the comb and rested his arms on the top railing.

I turned and leant on the railing beside him. 'Yeah, it can be a shit sometimes,' I said. We stared out to sea together.

BY NOW I'M halfway through the Brain Fitness program and I'm still noticing a gradual improvement. But I only manage thirty minutes of brain training per day; this is one limit I can't seem to budge. Maybe it's because I'm trying as hard as I can, and the program adjusts so that I'm always working at my mental capacity. Surely I must be one of the most motivated students Posit Science has had; it feels as if my life depends upon this working.

Yet it's not a consistent progression. If I have a day with lots of telephone calls or conversations with new people, my performance on all the brain exercises deteriorates. The same thing happens if I do any strenuous physical exercise.

I found out about the link between exercise and fatigue when I met up with Doctor Mercer on a Monday for my three-monthly review. As usual, she conducted the visual-field test first. When she printed off the result, it showed that I'd gone backwards. The black area in the visual field had increased since my last test; the previous one had shown a decrease. 'That doesn't seem right,' she said, looking puzzled.

I told her that I was feeling pretty flat, even though I hadn't done anything mentally taxing that day. I had rubber brain. As we discussed what might have brought it on, I realised that I was still physically fatigued from the competitive swim I'd done the day before. Doctor Mercer suggested a re-test a week later, when I had recovered. When I redid the test the next week, it showed that my vision was back to where it should've been.

The next weekend, at the surf club I ask Doug, 'Have you moved yet?'

He's been trying to sell his house, but he's not getting the offers he would like for it — the housing market's still depressed. The noise of his neighbours' music and dogs is irritating him like hell, and he wants to move into a quiet housing complex for those over fifty-five. 'I've been having trouble deciding what to do,' he tells me. 'I consider the options and then my brain freezes. You get to a point, your health improves, and then something comes along to set you back for a while.' He motions climbing down a ladder with his hands. 'But you don't fall back as low as you were before. You grow and learn from these experiences.'

He's right. It may not always feel like it, but the progress bar on the Brain Fitness program shows that I'm still ahead of where I started, even after a setback.

THE NEXT MONTH I meet with Kathy, the speech pathologist whom Leanne has organised. She will conduct my speech assessment over two two-hour consultations, a week apart. She tells me she has devised a non-standard assessment that will be sensitive to my deficits and will mirror the types of tasks I might do if I return to non-clinical research work; she has read the neuropsychologist's report.

In our first consultation, Kathy gives me a one-page summary of a physical-health study. It contains medical terminology that I'm not familiar with, but she thinks I'll work out the meaning from the context. After I've read it, I answer her questions about it.

Next, she asks me to correct errors of grammar and expression on the page, which requires me to look at the text again in detail. I do most of this successfully, but when Kathy asks me to recall the main finding of the study thirty minutes later, I confidently report the opposite result from the correct one: somewhere in that time, my mischievous brain has turned the most important bit of information from the study inside out.

At the end of our session, Kathy says that she has noticed my circumlocution. During the questions, she also observed that I needed to re-read the paper to get the answers, instead of recalling them from memory. This news is disheartening, but unsurprising.

In our second consultation, a week later, she gives me a ten-page research study to read. 'Evidence-Based Educational Guidelines for Stroke Survivors After Discharge Home' is much more challenging than the one-page summary. I understand most of what is written as I read it, but when I start a new section, I don't remember what I've read previously; conceptually, I'm not able to link the information from the different sections together in my mind. Kathy calls this a difficulty with 'integrating information'. And she notices that when answering her questions, I omit crucial information.

Kathy also says that I have word-retrieval problems with proper nouns: the names of things and people. She suggests that this is due to damage in my left posterior hippocampus. The hippocampus consolidates memories, so this makes sense to me. Proper nouns can't be replaced with another word, so no amount of circumlocution will help. Also, she says, I repeat things, as if I am telling them to her for the first time. And I often ask her, mid-reply, to say the question again.

Prior to my consultation with Kathy, my insurer sent me to see a vocational psychologist to find out what capacity I had for work, and to come up with ideas for a work trial. The psychologist sounded enthusiastic when we first spoke over the phone, promising to come up with options I hadn't thought of. The date of our appointment was my birthday, and this seemed a good sign. But at our meeting, I'd been thrown when she opened with: 'Tell me, where do you see yourself going with your future work?' I'd already given her my ideas over the phone and by email, and the thought of explaining it all again was deflating. I told her to ask me specific questions. I completed the questionnaires she gave me, and she promised to present me with her ideas after further research.

When her report came through, she had misspelt my name, got important dates wrong, and dismissed the suggestions I'd given her. Her ideas for a work trial did not take into account the cognitive difficulties I'd outlined. She recommended a trial answering telephone enquiries from the general public in the local-council office. This was a high-demand task for me because it relied solely on speech from people I couldn't see, and it required quick recall. As I didn't have previous knowledge of council services, I'd need to acquire a lot of new knowledge quickly.

Kathy, in the report she provides me with, says that I should avoid reception or directory tasks that rely heavily on auditory processing and the quick recall of information. I could undertake tasks that allow for delayed retrieval of information, such as writing. I would need flexibility with how, and the time I took, to produce the work. I would also need supervision from someone who could act as my editor, to help with the integration of information.

She also gives me a new insight: my good use of grammar and vocabulary masks my other cognitive deficits, giving a false impression of how well I'm functioning to others. Her conclusions confirm the feeling I'd had that the vocational psychologist was out of touch with brain impairment.

Kathy's results, like those of the neuropsychological assessment, validate my subjective experience. I am not imagining my difficulties; they are real. Her report becomes part of an arsenal of expert opinion for my insurer, treating practitioners, lawyers, and any doubters who want to query my incapacity.

When is a disability a disability? In the case of a hidden impediment, such as brain injury, I conclude that it is only a disability when others recognise the deficits. And it doesn't help that it has taken me fifteen months to work out what has changed since the stroke: it is often only in new situations that I learn what I can do as before and what I cannot. It's a meandering way to find out my limits, but there is no medical test that will work this out for me; life is the only test.

14

JAMES BENNETT-LEVY LEANS forward in his chair. With a look of concern, he asks, 'And how can I help *you*?' His spectacles concentrate the interest expressed by his eyes.

'I want to feel useful again,' I say. 'Be my prefrontal cortex; tell me what to do.' I explain my concentration limit for new things: two hours per day three days a week is probably all I can manage for the time being.

He nods in agreement.

After the debacle of the vocational psychologist's work recommendations and then Kathy's more informed report, I made contact with James Bennett-Levy, associate professor of mental health at the health-research unit that Leanne, the speech pathologist, had recommended to me. When we first spoke, James said that he remembered me from a positive review I wrote of one of his books, years earlier; I'd forgotten about this, but was glad for the connection.

Today he runs through a list of activities that he could use a volunteer to help with. They're all possibilities, although nothing grabs me. But my ears prick up at the last item on his list. He runs regular training programs in cognitive behaviour therapy, and has been asked by the university to design a psychotherapy course for mental-health professionals that incorporates the latest neuroscientific findings. Am I interested in neuroscience?

'I am,' I say. How serendipitous.

In fact, over the last few weeks, as my brain's improved, I've been doing a lot of reading on the subject. I've finished *The Brain That Changes Itself*, and it's given me new ways to reflect on my trauma experiences.

Norman Doidge says that the principle of neuroplasticity can be used not only to improve cognitive and motor function after brain injury, but also to aid in emotional recovery. If we can focus the mind away from worries or troubling images and instead dwell upon something pleasant, this rewires the brain for more positive associations — like a train being switched onto different railway tracks. I interpret this to mean that if I'm in a ruminating mood, I should distract myself with an activity that takes me out of my rumination. It's not important whether this is gardening, a coffee with friends, reading, swimming, or playing music; it's just important to do *something*.

Doug has read Doidge's book too, and he says that he's learnt to visualise his mind as an old-style telephone exchange. 'If I can see that a thought is not good for me, I pull that one out,' he says, demonstrating with his hands, 'and plug in a flower.'

I'm still doing psychotherapy with Wayne, and I've realised that the therapeutic progress I'm making has a neurological dimension. Traumatic memories are scorched into my brain because of the sensory and emotional impact of the original experience. In addition, when I recall a traumatic memory, with its emotional tone and patchwork of sensory associations, I'm reinforcing the memory's neural network: reimagining activates the same network all over again, etching the memory trace deeper, further sensitising me to potential threats. Wayne has reminded me that the threats I faced in the past are no longer present, so I no longer need to react. Yet my brain acts as if it doesn't know this: it still behaves as though those threats are real. In my hypervigilant state, my threat system is easily triggered: the demands of creditors, the children being difficult, and even a stern look from Anna are all enough to set it off sometimes. Present-day threats — or what I perceive as threats — set me off more than they should, even when I know that they have no direct connection with past trauma experiences.

When I can see that my emotional response to a past threat is outdated, new neural connections are made: an alternative network is created. As with learning any new skill, I need to keep activating the alternative network, separating the present from the past. For consolidation, this neuroplastic change requires that the connections in the older neural network weaken as the connections in the alternative network strengthen. In time, I will respond to what is *actually* happening around me, rather than what I imagine is happening.

Doidge refers to the work of neuropsychiatrist Eric Kandel. I remember excitedly reading an article a few years before in which Kandel was cited as saying that neuroscience has demonstrated that psychotherapy ('talk therapy') leads to physically observable, positive changes in the brain. When we re-experience trauma memories and uncontrollable emotions, blood flow to the prefrontal cortex is reduced, meaning that our threat-based brain structures take over. The neurological aim, therefore, is to activate the prefrontal cortex as much as possible before, during, and after these experiences. Talk therapy restores the prefrontal cortex's involvement in emotional response. As I read Doidge's description of this, I picture cerebral blood flow seesawing throughout my brain, as my prefrontal cortex and my limbic system wrestle with each other.

I also find out that psychological trauma causes the hippocampus to shrink, through the release of glucocorticoids (a class of hormones), especially cortisol. Prolonged release of cortisol, which occurs with depression and chronic stress, destroys cells in the hippocampus, causing memory loss. As people recover from depression, their hippocampal neurons can grow back, improving memory. So it's likely that even before the stroke, the ongoing financial stress, the post-traumatic stress disorder, and the depression caused shrinkage in my hippocampi; this is what the neuropsychologist was getting at. I wasn't imagining the memory and decision-making difficulties I had: there was a biological basis for them. Given that part of my left hippocampus was damaged in the stroke, it's fortunate for me that the hippocampus is one area of the brain where new neurons can grow from stem cells.

Antidepressant medications and St John's wort — and possibly psychotherapy too — can also increase the number of stem cells in the hippocampus. The creation of new neurons may be the indirect reason that antidepressant medication increases serotonin levels. Neuronal growth in the hippocampus improves a person's capacity to learn and remember, and to re-engage.

This gives me great hope. When James mentions his interest in the application of neuroscience to psychotherapy, I am already primed, keen to assist him in bringing these findings into the hands of other therapists.

After James sorts out the administration details that allow me to become an unpaid member of the research unit, he suggests a program of readings. We will meet up each month to discuss and summarise our findings. I am to concentrate on the authors Daniel Siegel, Louis Cozolino, and John Arden initially, reading their books to get an overview of the area of neuroscience and psychotherapy. Then, from time to time he will give me specific research articles to go through. Best of all, I can do this work at home, at my own pace. I'll need to make summary notes of what I read as I go, I tell him, or I'll forget it all. 'That would be great,' he says.

In fact, I've already started on Daniel Siegel's *Mindsight*. Siegel is a psychiatrist and an educator who pulls together neuroscience, psychotherapy, mindfulness, and case histories in a fluent way I've not come across before. My meeting with James has given me the impetus to finish it quickly. Soon, I move on to his earlier book *The Mindful Brain*.

What I learn here astounds me: Siegel says that mindfulness meditation changes the structure and function of the brain for the better. The key area is the prefrontal cortex (PFC), left and right. This is the foremost area of our brain — part of the frontal lobe — and it sits behind our eyes and forehead. The PFC is the thinking part of the brain; it gives humans the capabilities that most distinguish us from all other animals. It is divided into different areas: the medial PFC, the dorsolateral PFC, the ventrolateral PFC, the ventromedial PFC, and the orbitofrontal cortex. In those spending time practising mindfulness meditation, two parts of the brain — the medial prefrontal area on both sides, and the insula, especially that on the right side of the brain — thicken. Siegel mentions the orbitofrontal cortex, the anterior cingulate cortex, and the ventromedial PFC as the most affected areas.

Siegel cites the work of neuroscientist Richard Davidson, who found a shift in brain function to the left frontal lobe when people experienced positive emotions, those associated with facing challenges; and conversely, a shift to right frontal dominance when they experienced negative emotions, the type associated with avoidance. This left-brain shift also correlated with improved immune function and physical health. Davidson and others have observed that during meditation, practitioners show increased activation of the left PFC and less of the brain activity associated with mind-wandering: random thinking. They also exhibit greater brainwave synchrony — a finding consistent with better mental health.

The right hemisphere has a more direct link with raw sensory experience than the left hemisphere, and it processes distress and uncomfortable emotions. The left hemisphere has a narrator function that tries to make sense of our life experiences, and likes to put names to these experiences. But our autobiographical memories, in a non-verbal form, are located primarily in the right hemisphere. Mindfulness practice makes it more likely that a person will approach difficult situations (a behaviour associated with the left side) rather than avoid them (a behaviour associated with the right side). In addition, it stimulates the brain in ways that promote growth in the cortical regions that control emotions.

So I have a picture of a brain gym: I need to activate the left PFC to get me in approach mode and to get the different regions of the brain working together, allowing it to become unstuck from overriding emotions. Writing down my experiences and talking them through with Wayne are ways of making narratives, of making sense of events and activating my left PFC.

But I wonder, can I, or someone in a similar situation, find a way to not overreact, to stop seeing threats that aren't there? Non-reactivity, I find out, can be thought of as a way of pausing before making a response. And the key for doing this is strengthening the control the PFC has over the limbic system — the amygdala, the thalamus, the hypothalamic-pituitary-adrenal (HPA) axis, and the inner surfaces of the cerebral cortex. The limbic system is implicated in unconscious emotions: impulses such as sex, anger, pleasure, and the threat response.

The research tells me that mindfulness changes the connections between the PFC and the limbic system, tipping the balance more in favour of the PFC. The stronger the PFC's control — the left's especially — the more chance there is of a pause before responding to a limbic-system alert.

Meditation trains the mind to hold steady. It enhances the function of the anterior cingulate cortex. This cortex is an important link between the limbic system and the PFC, critical in snap decision-making, and, therefore, in the regulation of emotional arousal. Mindfulness is the microscope for seeing, or being in touch with, inner experiences. Through mind training, meditation is the dial for turning up the magnification.

Siegel says that there is no definitive explanation as to how focusing attention in the present moment, as in mindfulness, activates neural circuits in the way it does and leads to physiological, psychological, and interpersonal wellbeing. But I learn that mental noting, or the labelling of inner experiences, is a mindfulness practice that increases non-reactivity. People who use words to describe their internal states are more flexible and more able to regulate their emotions. Noting 'I am angry' or 'I am hungry' is a left-sided approach: it's a way of acknowledging an internal state, a form of approach rather than of avoidance. The act of labelling feelings is associated with reduced blood flow in the left and right amygdala, and increased blood flow in the prefrontal cortex — a physiological sign of greater PFC activation.

Siegel says something that particularly strikes me. We all experience the pain of loss, disappointment, and death — these are universal events. But our minds create mental anguish by grasping onto conceptualisations and reacting automatically. When our minds seize onto preconceived ideas of how something should be, rather than how it is, it creates tension. Instead, mindful moment-to-moment awareness that does not fix on judgements is a way of attuning the physical and psychological experiences of reality.

AS I CONTINUE with my reading and my meditation practice, I think I'm getting a better handle on the concept of mindfulness. Mindfulness is remembering to be aware of what I'm doing while I'm doing it, noticing what I'm thinking when I'm thinking it, noting what I'm feeling when I'm feeling it. This is creating the pause that I need to facilitate non-reactive actions.

It is my amygdala that sends me into a spin of fear over something small: the telephone ringing during business hours, a knock at the front door (our friends and family always come to the side entrance), an official-looking letter arriving in the post. I recognise my fear response as a feeling of dread: my body quivers with alertness; even the pores in my skin sense danger in the air. The dread tells me that something bad is about to happen. The feeling can last for hours or days. I wonder what is around the corner, what intention the stranger walking towards me on the footpath has, whether an unexpectedly loud sound or the shout of someone across the street means that something terrible has happened. The dread is malevolent; it wants to destroy my family and me.

But if the threat is like a ghostly presence that never solidifies; if I can't surmise its shape, texture, or colour; if I can't work out if it is large or small, quick or slow, how do I respond to it?

In the mornings, there is a peewee out the back of our house. It perches on the end of the clothesline, peering at the door of the garage. It's a black-and-white bird, a small version of a magpie, with a tiny face and a determined beak. It attacks the windowpane in the door, repeatedly flying into it — backwards and forwards from the clothesline. It must see its own reflection in the window and think there is another peewee in its territory.

Bang! Bang!

Surely this must hurt?

I sit in the downstairs office each morning and hear it, day after day, railing against its hollow enemy, while I rail against mine.

I am beginning to see that the threat is internal as much as external. Yes, there are external threats — in particular, the prospect of bankruptcy and what this will mean for the family. But no one is going to kill us. No one is harming us, even if it feels this way.

The cortical response and the limbic response are two different pathways. The limbic response is unconscious; it works at lightning speed, makes gross generalisations, and activates the body immediately for fight, flight, or freeze. Some call it the 'low road'. The low road is what saves us in an emergency, compelling us to act immediately and ask questions later. The hypothalamus and the pituitary and adrenal glands — the HPA axis — are the workhorses of the fight-or-flight response. The HPA axis causes the pupils to dilate, the heart to beat faster and more strongly, breathing to quicken, stomach and bowel activity to slow, the bladder to constrict, and blood pressure to rise.

The cortical response, directed by the prefrontal cortex, is the conscious voice of reason. Some call it the 'high road'. The high road is like the parent who intervenes in a squabble between two siblings — quieting them sufficiently to hear what the squabble is about and then deciding what is to be done.

This takes time.

My blow-up with Anna and the children prior to my stroke, when they were watching television, was my 'low road' in action. The subcortical region of the brain took over my body and I temporarily lost rational control. The family weren't the real threat; they were the trigger. My amygdalae were already in an overheated state from the stressors of the day, and, with my mental fatigue, it didn't take much to tip me into the fight response: rage. So I couldn't help my initial 'low road' reaction. It was when my PFC came online again that I regained conscious control — my humanness. The 'low road' might never have been activated if I had been functioning normally.

The neural pathway from the amygdala to the PFC is thicker, and therefore transmits information faster, than the pathway going from the PFC to the amygdala. This enables the amygdala, with the limbic system, to dominate the PFC for a time. If the PFC decides that the threat is not real, or has passed, it cools the amygdala's response through the release of the neurotransmitter gamma-aminobutyric acid (GABA). Oxytocin, a neurohormone associated with trust and safety, also 'cools' the amygdala. And more serotonin in the limbic system means less of the fight-or-flight response.

Pleasurable social activities, such as watching an enjoyable movie, being with friends and caring people, playing music, and swimming in the ocean, increase serotonin and oxytocin levels. These activities also take my conscious attention away from internal triggers: bodily sensations, thoughts, and memories that elicit dread. Physical activity uses up the adrenaline in my body, just as the fight-or-flight response was designed to do.

LATER THAT OCTOBER, James and I attend a two-day professional workshop on the brain and anxiety, given by a clinical psychologist. It's the first such workshop I've come across. James is keen to see how someone else teaches this stuff.

It's held in a large metropolitan hospital, and I've booked two nights' accommodation within walking distance — the prospect of driving in the city is too frightening, and I want to conserve my mental energy. Getting out of the lift, I bump into an acquaintance, causing us both to do a double take. I learn that he's here for treatment of a substance addiction; he's equally surprised to find out that I'm here for my brain.

After registration, I enter the conference room: there are easily more than one hundred people sitting in closely spaced chairs. I can't see James, and I scramble over bags and umbrellas spilling around people's feet to get to a vacant chair in one of the rows towards the back. Brain science and anxiety is clearly a hot topic.

The psychologist, Pieter, reminds us that psychotherapy changes the brain, and that understanding the brain's processes will make us more patient with the rate of psychological change. He says that a healthy-functioning left PFC is required for good mental health in most people. Usually the left PFC is able to inhibit the activity of the right, and information flows from the right PFC to the left.

The left PFC's functions include categorising, problem-solving, detailing, and rational analysis. It puts events in order, and in time and place. It is primarily involved in verbal work and in making meaning of events. The right PFC specialises in non-verbal recognition and emotional memory. It recognises faces, reads others' emotions, and assesses the emotional significance of events. It is involved in creative, non-verbal problem-solving, and in spatial relationships, such as rhythm in speech, music, and movement. It has a primary role in the processing of sensory data.

We learn that during trauma and the re-experiencing of trauma memories, blood flow to the left PFC temporarily drops relative to the blood flow in the right PFC; the left PFC's influence diminishes, while the right PFC dominates. Anxious thoughts and emotions become 'stuck' in the right hemisphere when cut off from the left's means of processing. I remember the assertion in *Mindfulness-Based Cognitive Therapy for Depression* that you 'can't think your way out of depression'. Depression requires action and not avoidance to get through it. Trauma causes hypersensitivity to the cues of trauma: anything that might remind the person of the original trauma. For example, seeing smoke rising in the air became a cue for my ambulance-driver client, who was traumatised after seeing incinerated bodies in a car crash.

Pieter gives us an explanation that shows why my dread has such a hold on me. He says that fear drives the mental error that the symptoms of fear are evidence of a serious threat, when in fact there is no real threat. The more the brain goes into panic, the more sensitised it becomes to future attacks of fear. He confirms what I have taken from my reading of Dan Siegel and others: if I am to get well and get my brain back in balance, I need to activate my left PFC more. I need to reduce the physiological triggers causing the fight-or-flight response.

Pieter says that activation of the left PFC cannot be done chemically, but he gives a number of strategies to do it. Exercise dissipates adrenaline, noradrenaline, and cortisol in the body, and uses up the energy released in the stress response. It raises the level of serotonin and endorphins, promoting a positive mood. He recommends a brisk walk of up to forty-five minutes five to seven times a week. I put up my hand and ask about yoga and Pilates. Pieter says that these activities do not release the same level of endorphins as exercise, but they do promote mindfulness of the body. Yes, I've noticed this, with Pilates especially.

Good-quality sleep is also critical for brain function, he says. Brainwave patterns allow the hippocampus to send information to the left PFC and into long-term memory. Rapid eye movement (REM) sleep relieves stress. During REM, more neurotransmitters are produced, through the process of regeneration. I remember that Norman Doidge mentions something similar in his book: he states that sleep, and REM sleep in particular, facilitates neuroplastic change, as it assists the growth of new neurons and of myelin.

During sleep, the brain sifts through the short-term memories of day-to-day events, consolidating the important ones into long-term memories. So I need to do whatever I can to sleep better. Wayne has already told me this too; now I've got further confirmation.

According to Pieter, changing the pattern of fearful thoughts requires mindfulness: conscious awareness of thoughts and sensations in the body. Writing about personal experiences activates the left PFC and can create a sense of safety and control. The writing process increases blood flow in the left PFC. I remember how writing out a list of trauma memories the day before the stroke was the only way I could get them 'out' of my head.

Art therapy reduces thought loops in the right PFC by allowing the expression of emotions in a non-verbal form. Music activates the left PFC by forcing the brain to think in sequences and to remember; listening to music is less effective than making it, I learn. Conversing with a friend activates the left PFC. Travel and walking also enhance neuroplasticity, Pieter says.

Novelty — being in new situations or learning new skills — challenges the brain, keeping existing neurons alive. It is the old catch-cry of 'use it or lose it'. Novelty promotes new neural connections, stimulates the growth of myelin, and can trigger neurogenesis. The more concentration it involves, the more neuroplastic change it induces. I think of my music: singing lessons with Lily, duets with Nick. These are new experiences and new skills, challenging my brain. I'm heartened to learn how much I'm doing right.

Someone asks how long it takes for brain changes to occur with therapy. Pieter tells us that with psychotherapy, consisting of one or two weekly sessions and daily homework exercises, it will take about four months for the therapeutic change and new neural pathways to become established. He says that psychological resistance can be thought of as a reversion to the old neural pathway.

The course is hard-going: I have a long sleep during lunchtime on the first day, which re-energises me for the afternoon. On the second day, after lunch, I can feel rubber brain setting in, but I want to stay on because it is so valuable. Yet by the mid-afternoon tea break, it's clear that I'm not taking anything in. I catch up with James and tell him that I'm leaving; he'll get a copy of any new notes for me and we'll chat about the workshop at our next meeting.

I realise after the course that one thing Pieter didn't address specifically is food for the brain. Over the next few days, as I'm reviewing my notes on the course, I consult various articles. I learn that the omega-3 fatty acids found in fish oil promote neuronal growth, improve mood, and slow cognitive decline. I read that vitamin B is critical in the synthesis of many neurotransmitters, and that vitamin E is the main antioxidant in the cellular membranes of the brain. I think I get enough Vitamin E in my diet, but I'm not sure about B.

My newest treatment practitioner is Doctor Franklin. On my fourth session with Doctor Banister, after the stroke diagnosis had been confirmed, he had seemed preoccupied and said he needed to make a home visit straight after our interview. He offered me few useful insights or advice. He was the most expensive doctor I'd seen, and I had difficulty paying his fees. I had never liked having a client of mine terminate sessions with me, but I was not happy with the service Doctor Banister was providing. I spoke with a GP acquaintance I bumped into, and she encouraged me to change psychiatrists, saying that I did not need to give a reason. I decided to move to Doctor Franklin, who has a private practice. (In our last session together, Doctor Banister said I could return to treating anxiety clients, but not traumatised ones; he was really out of touch with what was happening with me.)

So on my next visit to Doctor Franklin, I ask about vitamins. He encourages me to take omega-3, but insists that I take 1000 milligrams of the DHA and EPA combination per day. He explains how to read the information panel on the side of the bottle to assess the percentage of these fatty acids in each capsule, so I can determine how many I need to take to reach 1000 milligrams. He mentions vitamin B12 as especially important for me because I do not have a high-meat diet, although I have an adequate intake of fruit and vegetables. I've still been taking 100 milligrams of aspirin daily as a blood thinner, on Doctor Small's advice, and the St John's wort. So each morning and evening, I set up my collection of pills on the kitchen bench, pleased that I seem to have all bases covered.

15

THROUGH MY READING and discussions with James, I learn that we possess a system of mirror neurons, as part of the motor cortex, which allows us to interpret another person's actions. Mirror neurons fire both when we act and when we observe someone else performing the same action. For example, if I see someone lift a cup in their hand, I can anticipate that they are going to drink from it because the same neurons that would be involved if I were thinking of taking a sip are firing in my brain.

Mirror neurons are also found in the frontal and parietal lobes, enabling humans to gauge another person's intentions and sensory experience. James and I are especially interested in the part the mirror-neuron system plays in therapeutic empathy.

Some locations in the brain are at the centre of several neural pathways, like busy crossroads. The insula is one of these. It is part of the cerebral cortex, and at the crossroads between the mirror-neuron system and the subcortical areas of the brain and the frontal lobe. Sensory information comes to the cortex from throughout our body — from the major organs such as the lungs, the heart, and the intestines, and from our peripheries — via the insula.

The mirror-neuron system is also part of a network that gives us the capacity to infer others' emotional states. Our feelings can originate from our bodily organs, the brain stem, and the limbic area, and the input these provide to cortical functioning. This is how a 'gut feeling' or 'intuition' is formed. The insula appears to be the nub of our interoceptive sense — the awareness of our internal states.

Siegel calls the neural pathways that let us read others' intentions and feelings the 'resonance circuits'. The insula and the mirror-neuron system are part of these. How well we intuit another person's state of mind depends upon how well we know our own. We use our bodies to resonate with the emotions of others. People who are more aware of their internal states can be more empathic with others.

Sensory experiences are encoded in our neural networks, so they have the potential to be recollected as memories. Scientists distinguish two types of memories: implicit and explicit. Implicit memories form without conscious focused attention. When we recall an implicit memory, we don't have the sense of bringing it up from the past. This is because the hippocampus is less involved in the formation and recall of implicit memories. Procedural memory, which is involved in riding a bicycle or driving a car, is a type of implicit memory. We do such things without needing to recall how and when we learnt them. Phobias can be underpinned by implicit memories. A phobia of dogs — an emotion-laden implicit memory — can be caused by an early childhood experience of being frightened by a dog, and yet the person with the phobia often has no conscious memory of this happening.

Implicit memories colour our psychological states without us being conscious of why. When we meet someone for the first time, we have an automatic, unconscious reaction to them, whether negative, positive, or neutral. This response is based on past experiences we've had of other people. I dislike 'smelly fish', and when I told my sister of this once, she reminded me that our mother fed us tin after tin of sardines when we were young: something I had forgotten. My sister doesn't like smelly fish either.

An explicit memory needs to be consciously recalled or remembered. It can be a factual memory (semantic memory) or a memory of a life event (episodic memory). For example, I remember specific occasions when my father or a driving instructor gave me driving lessons: episodic memory.

Memories, with their sensory and emotional components, are stored throughout the brain. But it is the hippocampi's role to encode and recollect explicit memories. The left hippocampus specialises in facts, while the right specialises in self-related memory. If the hippocampus is temporarily shut down during an experience, a person's memory of that event is fractured. Rage, terror, and other intense emotions, as well as alcohol, drugs, and neurological illness, can shut down the hippocampus.

I realise now that my amnesia following the stroke was contributed to by my left hippocampus being partially disabled, leading me to think my first day in hospital was only an hour long.

In a flashback, as occurs in post-traumatic stress disorder, only fragments of images, sounds, touch, and smells are recalled, but the emotional re-experience can be as terrifying as the original. It feels as if it is happening all over again, not as if it is simply a memory from the past. The flashback is triggered by perceptions, emotions, and actions that bear some resemblance to the original event. Recurring flashbacks strengthen the memory.

Psychotherapy tries to bring the hippocampus and the explicit-memory system into play. This creates new neural connections. Therapy requires a dual focus of attention: the person re-experiences the traumatic memory while also being aware that they are in the present moment where nothing bad is happening.

Psychotherapy needs to be done in a safe place, with a person the client trusts, and sees as skilled and protective. The individual needs to re-experience memories while being aware that they are safe, and not in the same time and place as the original traumatic experience.

Wayne is the mental-health practitioner with whom I feel the safest, followed by Doctor Small and Doctor Franklin. Before my stroke, in early 2009, Wayne and I had gone through a process of desensitising individual trauma memories. He'd thought that I had them under control before our financial situation blew up. Yet the demands and threats from creditors had re-activated them and they were re-traumatising me; I couldn't deal effectively with the financial threats in my current state, he said.

He had explained the desensitising process while I sat on the sofa in his office. 'I will remind you that these are memories of the past, and I will be asking you to look at them as the person you are now, with all your current strengths. We will be working with your intuition, which brings up material that needs airing.

'I would like you to close your eyes and show me a sign — perhaps raise a finger to indicate "yes" — so I know when you are ready to proceed without you needing to speak to me,' he continued.

I raised my left index finger.

'I would like you to go back through your past, bringing to mind a troubling memory, and when you're ready, describe it to me.'

Flashes of memory emerged out of darkness, filling my mind's eye like a slideshow and then disappearing, until I was left with one that stuck. It was of a time when I was working in the prison system. I had just run the morning group-therapy session in the Drug and Alcohol Rehabilitation Unit, which housed about eight inmates. I usually conducted individual counselling sessions with one or two of the inmates after the group. This particular day, Rob, an armed robber with a drug addiction, sat across from me at a small table in the corridor around the corner from the group room. He was a muscular man, clean-shaven, with a crew cut and tatts on both arms. Dressed in green shorts and a tan-coloured singlet, he exuded an air of angry indignation.

I was, theoretically, under the observation of the prison officers on the other side of the glass security screen that looked into the main room, with one officer present in the group room. But if an inmate became violent, the worst would've been over by the time the officers could give assistance.

Rob had been telling me he had no choice in committing the crimes of which he had been convicted.

'But you did have a choice,' I said.

This set off an internal spring in him. 'I fuckin' didn't,' he said, jumping to his feet. He leant over the table, staring me down, his face pale with rage, his right elbow cocked, ready to pummel me.

I threw my arm across my face and leant back in my chair. We held our poses for what seemed like minutes, and then, all of a sudden, his face transformed, as if a demon had left it, and he sat back down. His gaze rested on the table. Nothing was said. No officer came over to us.

I was shaken, but told myself that I hadn't been hurt. 'Want to start again?' I said.

He nodded, and our conversation sputtered back into life.

At the time, I had seen this as a normal part of my job — something I should be able to deal with. But I was surprised, sitting in Wayne's office, how insistent this memory was — how alive it was — more than twenty years later.

My eyes were still closed when Wayne asked, 'Would it be okay to look at this memory from the perspective of who you are now, with all your current strengths?'

'Yes,' I said.

'What would you like to say to your younger self, to reassure him?'

I pictured my younger self: long-haired, bearded, a man with a passion to help others. I said, 'In the end you were not hurt. You did not deserve to be threatened in that way. You were doing your best and working in a difficult environment. You did okay. You have nothing to be ashamed of.'

My younger self was eager for reassurance from an elder. He didn't have an available senior with whom he could debrief. So, more than twenty years later, the words that needed to be said then had been said.

Wayne said, 'Would you like to give your younger self a hug, perhaps?'

He looked so vulnerable, so alone. I gave him a good hug. I felt such tenderness for him. I understood now that he'd taken on the armour of invincibility that is part of youth's skin, treating life's insults like irritants.

I realise, after all I've learnt about the brain, that when Wayne had me go back to this memory, he was pairing the re-experiencing of this incident with my adult self, with the wisdom and strengths I now have.

The PFC gives us the capacity to think about thinking and conscious awareness. It enables us to plan, reflect, and learn, but it also enables worry and self-doubt. Was I using my PFC's capacity to override the visceral and emotional responses I was having to my work, expecting that I should cope with what was in front of me, as my mother had? Was I being dismissive of the distress smouldering within me?

As the weeks pass, I begin to put into practice new ways to engage the left PFC. Before the stroke, I'd stopped yoga and taken up Pilates after I'd injured my shoulder and couldn't bear weight on it. Also, the stroke had left me with low blood pressure and dizziness, and I realised that I needed to be active in ways that didn't involve moving my head up and down repeatedly — the experience at Nippers had taught me that. But Pilates is less strenuous, and it targets specific muscles and movements in my body, enhancing mindfulness.

I know that mindfulness and meditation soothes the arousal of the sympathetic nervous system. The more I can notice my thoughts, feelings, and bodily sensations, the better chance I will have to pause, and to respond to them differently. I try to keep reminding myself that when I experience dread, this response does not mean that I am really under attack.

Some time ago, Doctor Small had told me how much he enjoyed reading in his medical journals stories with people's firsthand descriptions of their health conditions. 'I remember more from these personal accounts than I do from yet another double-blind research study,' he'd said. He had encouraged me to write about my experiences with my stroke. Now I know that writing about what has happened to me, and is happening to me, will activate my left PFC. It will unstick the ruminating thoughts and emotions and help me to make sense of what has occurred.

IN LATE OCTOBER, we make a trip to our favourite national park for a two-day stay in the same guesthouse we stayed in when my sister called to give me the news that Dad was ill. This time, we're joined by another family and their two children. Their children are best friends with our youngest two.

On the first morning, we talk through our plans for the day ahead. The other parents want to do a short walk. I'm keen to make the most of this time in nature; I want us to go on a daylong hike. The children are capable of doing this, so I don't see what the problem is. As we talk further, I become aware of my annoyance: a certain tightness in my body and a narrowing of my thinking.

Then something happens: I allow myself to let go of the need to determine the outcome, to have it go my way; I soften my mental grip. Then it's easy to accept a neat compromise: the other family, their two children, and our youngest two will go on the shorter walk. I, Anna, Ashley, and her friend, who has also come with us, will go on the long walk. Solved.

I can't help but feel that the mindfulness training is kicking in and my PFC is being more assertive. It's working.

IF ONLY THE momentum of life went in one direction, but it's wayward. The next month, things begin to slide backwards, and again I'm caught out.

Sixteen months post-stroke, the aspect that still surprises me the most is the mental fatigue I get after concentrating for a few hours. When I go past my limit for a few days in a row, it gives me a feeling of being really, really worn out, as if I need a holiday. It's become clear to me in the last few months that the fatigue is worsened by physical exhaustion, insufficient sleep, and exposure to loud noise, large groups, new people, and stress.

In late November 2010, I awake early one school morning with rubber brain, but I get up and meditate in the cubby house anyway. As I try to focus on my breath, the home phone keeps ringing, but none of the girls answer it. Who's calling, and why won't someone pick up the phone to make the noise stop?

Later, after breakfast, I feel rushed and unsettled as I make the school lunches in the kitchen. Anna has already left, so I will be getting the girls off to school on my own. The dread is with me, and it intensifies by the minute. I butter crackers, but each one I pick up breaks, my hands are shaking so much.

I can't do this. I can't cope anymore.

Impulsively, I throw my favourite cereal bowl into the sink, followed by my tea mug. The loud crash it makes is oddly exhilarating. I pick up the plastic box holding the cartons of crackers and propel it into the pantry door, as if throwing a football. The jar of Vegemite follows. A cascade of foodstuffs clatters off the shelves.

I've been trying to hold everything together, but as the year has gone on, my reservoir of fatigue has risen like floodwaters.

I phone Anna in desperation. 'I can't cope anymore. You'd better come home now.' She tells me that she'd been the one ringing earlier.

I drive Ashley to the bus stop because she is running late again. The two younger ones are nowhere to be seen. When I come back, I find them in the kitchen, picking through the debris, commenting like disaster experts on the turn of events. The pantry floor is frothing with fallen boxes and jars, and crumbs from the broken crackers dot the ground like confetti. My cereal bowl lies in fragments in the sink.

'I'm not angry at you,' I say to the girls.

'Dad, having one of your three-year-old tantrums again?' Emma says, with a tentative smile.

I manage a little laugh. Her humour is a relief.

They help me to tidy up and then go out onto the verandah to have their toast. Anna arrives and goes out to sit with them.

I'm still incapable of making the lunches, so I join them. Not a lot is said. When the girls have finished, Anna offers to take them to school. I farewell each child with a hug. 'I'm not angry with you, you know that.'

'Yes,' Amelia says. She seems unperturbed.

Emma says, 'Yes, but it scares me.'

The rest of the day is a blur — it's hard to concentrate on anything when I have sore brain. I pick the children up after school, and they don't appear annoyed with me. Later Ashley joins me for swim squad, and after dinner we watch *Julie and Julia* — Amelia's pick.

The next day, Anna tells me that she needs to talk. Hesitantly, she says she may be suffering from compassion fatigue over my condition. I tell her that I am not coping with the normal daily demands along with my rehabilitation commitments. Yesterday's incident has shown me I must change something, that I need a rest. I'm worried about Anna's health too; I think she needs to sleep and to exercise more.

I know that I've been reaching my limit more and more easily recently. I think what's tipping me over the edge is the work trial with James, and the extra mental fatigue this creates. My insurer has contracted a local rehabilitation manager, Matt, to draw up a rehabilitation plan for me. It includes the work with James and sessions with an exercise physiologist, as well as a new laptop, software, and ergonomic accessories for my work desk at home. They are all supposed to help me, but learning how to use them creates an added pressure. We are also looking at vocational options that don't involve me going back to clinical work.

Matt has worked with many brain-injured clients in the past, and has told me that mental fatigue is a frequent and long-lasting outcome of brain injury. He says that I need to be proactive in managing my fatigue: to anticipate demands and situations that will tire me out, resting beforehand and slotting in recovery time afterwards. But what I can't anticipate is the unexpected and urgent demands from real-estate agents, lawyers, and anyone involved in the sale and management of our properties — few of the sales have been straightforward or easy. I usually run the children around to their after-school activities: swimming classes, water polo, little athletics, dental and medical appointments. Then there are the days you can't predict, like one daughter losing her hearing aid (she has moderate hearing loss in one ear) and spilling superglue onto her jeans, causing them to stick to her leg. The only way to remove the jeans was to cut them off her, and then she sat in a warm soapy bath until the stuck-on patch loosened.

My fatigue worsens on the weekends, with physical exertion and the cumulative effect of the noise and demands of the children. On Mondays, I often have a hangover-like effect until Tuesday morning, when my brain finally allows me to be productive again, and I can start on James's work.

Anna thinks it would be good for me to go away for a time, so we can all have a break. 'India, or to study a course,' she says. To help with the daily demands, she suggests the children make their own lunches and that Ashley get herself ready early enough to walk to the bus stop.

Her suggestion to go away plants a seed in me. Even though I'm making progress, I haven't got there yet. Time out might be just what I need right now.

16

THE RAIN THUDS on the car roof as the wipers shift the deluge from the windscreen. Overtaking vehicles send curtains of water at me. It is 6 January 2011, and I'm heading north, to Hervey Bay.

Before Christmas, I decided that a meditation retreat in a place of silence and calm, without the family, would be the ideal spot for me to recuperate from my accumulated fatigue. But I couldn't find anything to fit my dates and budget. So I phoned Choeying and asked if I could come and stay with her.

'Yes, of course, darling,' she said, sounding delighted. 'How's that for karma in action for you: there is another brain-injured person coming to stay in January. You can join us for the calm-abiding course I'll be facilitating.'

In the middle of last year, there was a call-out in our community for billets for people attending a teaching given by a visiting international Buddhist teacher I'd not heard of before (Dzongsar Jamyang Khyentse Rinpoche). I hadn't intended on going, but I wanted to help out, as others had helped me to attend various courses. Anna agreed too.

When I'd picked up our assigned billet from the bus station, she turned out to be a Tibetan Buddhist nun, Choeying (her spiritual name, pronounced *Chur-ying*). After we talked for a while, I realised that she was the shaven-headed person in maroon and yellow standing at the front of the cordon in the hotel lobby, waiting to greet the Dalai Lama, the same day I met him.

Over the course of the nine days that Choeying stayed with us, Anna and I each talked with her. I quickly sensed that I could trust her and could speak frankly. I told her how panicked I was about our finances; I was worried about the effect that bankruptcy would have on the children's lives. She said that great suffering could be a curse, or a blessing or gift, depending upon how we saw it — it offered the opportunity for change, to discover humility and self-compassion. If I managed it with grace and equanimity, the children would model that. They did not need to have everything given to them, or they would not learn gratitude and resilience.

I found her advice to see my present situation as a 'blessing or gift' a bit weird, and not overtly helpful. Still, like others I had gravitated towards — such as Wayne, Doug, and Doctor Small — I felt calm in her presence. Somehow, it tamed the dread. At the end of her stay Choeying invited us to visit her, and January seemed like the perfect time to take her up on it.

It's been a long drive, and by now it's late afternoon. The last of the daylight is fighting with the darkening sky. The raindrops become heavier, until they sound like tribal dancers on the roof. I will need to stop for a break soon.

I wonder about the wisdom of my decision to head north during the tropical summer season. Not long before, on the approach to a new town, I came over a rise in the road and was confronted with large lakes on either side of the highway, water nipping at the edges of the bitumen. A few more hours of this rain and I'll be cut off from home.

But I put these thoughts aside. Since the stroke, I've learnt to trust my intuition. My recovery has shown me that I need to go to uncomfortable places sometimes.

I've gone through a lot of discomfort in the last twelve months, but I've made progress. In early 2010, Anna and I made the heartbreaking decision to put our home on the market, but it took until late December 2010 to sell. The buyer couldn't move in for a few years, so now we're renting it back from them. Once the sale is settled, we can pay back the loans from our family and friends. We've sold the investment properties that were saleable and paid out their respective mortgages, leaving only those that won't achieve a sale price sufficient to pay out their loans. I'm gratified with what we've achieved; financially, we've got through a lot this year, although it's been about mitigating our losses rather than making any gains.

We're not out of hot water yet. The developer that commenced litigation against us went into administration a few weeks ago. But just before this, they drew on the deposit bond, which guarantees payment of 10 per cent of the purchase price. We don't have this kind of money. The finance company that issued the deposit bond sent us a letter of demand: pay them the full value of the deposit bond or else. I phoned the company, hoping that we could reach an agreement that took into account our actual financial position. I was passed from person to person before ending up with the man who made decisions about these things. I outlined the purpose of my call, but he refused to speak with me further, saying that we needed to go through a lawyer. So we'll have a new legal fight on our hands this coming year.

But I'm more hopeful this time around. We found a service that negotiates with creditors on behalf of their clients. Dominique, our lawyer there, understands: she had to claw her own way out of debt after failed property investments. She's made applications for financial hardship on our behalf with each of our creditors. Dominique has also put us in touch with the Financial Ombudsman Service, an independent body that mediates in financial disputes with lenders. After we lodged the disputes, the pummelling of creditors' demands stopped — at last giving us breathing space and time to think, and giving my fight-or-flight response a rest.

The scenery is changing; houses are showing through the trees beside the road. My GPS has woken up and is fussing over the route ahead. I've reached the outskirts of Hervey Bay.

Following the GPS instructions, I navigate the streets, finally pulling into a cul-de-sac. I see the place I'm looking for: an unusual red wooden fence painted with horizontal stripes; two large palm trees in the front garden; and, behind these, a two-storey maroon home. I nose the car into the driveway and pass a stone Buddha sitting alone in the middle of the lawn. Navigating the long driveway, I stop in front of a garage-cum-shed. I get out and walk back part of the way along the drive, poking my head in through a sliding glass door. I see a large room. To my left is a heavy bookcase filled with spiritual books; to my right is an elongated area with white mats and rectangular-shaped cushions on the floor, oriented to face a white armchair. Before the armchair sits a low table. This must be the meditation hall.

'Hello?' I call out.

A curtain is brushed aside, revealing a small office space off to the side, and a woman emerges. It is not Choeying. This woman is middle-aged with short hair, glasses, and sallow skin. She smiles. 'You must be David. We were expecting you. My name's Queenie.' Queenie explains that she's helping with the organisation of the calm-abiding course. 'You'll be attending, won't you?' she says.

'I guess so,' I say.

Queenie takes me upstairs to the kitchen and makes me a cup of tea. She will be staying for the whole month with Choeying and Rex, she says, while she detoxifies from alcohol. Quite unselfconsciously, she tells me that her usual routine was to start on the grog on Friday evenings and drink through the weekend. She abstained from alcohol for the rest of the week, but she still felt its after-effects, each time wondering why she had these weekly blowouts.

Choeying arrives. She's wearing her usual loose, white cotton garments, her hair very short and white. Enveloping me with her pale arms, she says, 'Hello, darling! So pleased you could come.'

She introduces me to Rex, her partner, who has just materialised. He's tall and broad-chested, bald with wire-rimmed glasses, giving the impression of a rather oversized, good-natured professor. He looks to be in his mid-sixties, and speaks with a gravelly, resonant voice.

The house is full of guests already, so Rex tells me I can sleep in the caravan under the carport, in front of the shed. We go out to inspect it, and it's delightfully retro: an orange-and-brown 1970s Millard with a single bed. It's only a short walk to the back staircase, which leads into the kitchen and dining area. It's private, and should be quiet; I'm going to like it here.

THE FOLLOWING MORNING, Saturday, after a good night's rest, I get up early and head into the meditation hall. Sitting on one of the cushions, I try to meditate, but I still have rubber brain from yesterday's drive. After half an hour I give up, and head upstairs for breakfast.

Rex is there. He's in a plain T-shirt and shorts, with bare feet. He starts work early, he tells me, and he's having his first break. Would I like a cup of tea?

'Yes, please.'

He stuffs the pot generously with leaves. 'I like a proper cup,' he says, grinning.

Once the tea is brewed, we sit across from each other at the long dining table. Rex slips into telling me about himself. He trained as an engineer and served in the army, achieving the rank of major just before he left. Since then, he's run several businesses and is now an inventor. The shed downstairs is where he works.

He is not a Buddhist, he says, but supports Choeying in her work. They have been together for more than twenty years. They've both had previous marriages, from which they have adult children.

Early in their relationship, Choeying told him there were things in her past that he should know about before he decided to commit to her. 'My initial thought was, "It's abuse," but I found out that it was serial childhood abuse — sexual, emotional, and physical. Around the time we met, she started to have nightmares and intense anxiety.

'I thought sexual abuse — the physical act — was an awful thing, but I found out that the worst thing was the grooming, the fear and guilt it created in its victims.'

They went on a driving holiday, at the recommendation of their GP. 'One night I woke up to find she was at the window of the hotel room, about to jump out onto the concrete below. I pretty much just managed to catch her by gripping her waistband with my fingers. Suicide attempts became a recurring nightmare. We ended up in the Royal Women's Hospital in Melbourne. They sent her to a psychiatrist, who conducted tests and recommended psychotherapy. We sold the marketing business Choeying had spent eight years setting up and focused on getting her well.

'Psychotherapy affirmed that something bad had happened, shouldn't have happened, and that she was not guilty for it happening. But Buddhism did more for her than psychotherapy.'

I'm intrigued, and ask what he means.

'Just as she was becoming stable and our life was getting back to normal, she got cancer. This must've been around 2002 or 2003. It was a rare form of cancer in the roof of her mouth, and aggressive. She was experiencing massive head pains due to it, but because she had a history of mental disorder, the doctors thought these pains were psychogenic in origin and diagnosed trigeminal neuralgia.' This is a condition, I know from my pain-management work, that causes extreme facial pain. 'It wasn't until they did MRIs that they made the correct diagnosis.

'Following the operation to remove the growth from her mouth, a biopsy revealed that it was a more serious condition. She was rushed into an emergency operation, and radiation therapy followed. For the next six weeks, she was given radiotherapy every day. She had claustrophobia, but was able to go into MRI machines. It involved her being in a confined space, unable to move, with her head secured. Bloody awful. She would stick a picture of the Medicine Buddha up above her in the machine during treatment, so that she could imagine the radiation rays were the healing light from the Buddha. Meditation carried her through.

'She learnt that all these things are surmountable. She doesn't want everything she's been through to be a waste. She wants to help others dig themselves out of a hole, even when the holes may not be as big as hers was. I understand the motivation to be useful, as I felt the same in providing support to our soldiers as an army engineer during the Vietnam War.

'Choeying's teacher, Robina, an Australian Tibetan Buddhist, was running the Australian arm of the Liberation Prison Project, and she asked us to take over when she moved to America. There was a little group going into Rockhampton Prison, and they kept asking for someone to go up there. Although it involved a seven-hour drive, Choeying was prepared to go. Also, at the house here, a few local people started to come round, and this developed into a sizeable weekly meditation group.'

It's an unusual story, but I can believe it. It's not like an ex–army major and engineer to make up this sort of stuff. Choeying has hinted at a difficult childhood to me before, and mentioned her recent recovery from cancer. I know from my professional work that someone who pulls through this kind of personal history and thrives cannot remain ordinary. I feel lucky to be staying with Choeying and Rex.

ON SUNDAY MORNING the first session of the calm-abiding course begins, and I join in. The course is being run over the coming three weeks, with a mid-week evening session and an all-day session each weekend. There are twelve students attending.

Choeying sits at the end of the meditation hall on the white chair, on a meditation cushion, with her notes propped up in front of her. She begins by welcoming us all, speaking in a commanding, no-nonsense manner. She was a businesswoman, she tells us, before becoming undone. The 'unimaginable abuse' she had experienced as a child caught up with her and she became suicidal. She has overcome cancer twice in her life — the first time in her twenties, and the second more recently, with Rex by her side. Her adult daughter is estranged from her. 'So darlings, there is nothing you can tell me that will shock me, that I haven't experienced or heard before.'

She asks us to say our names and why we've come. A big man, ruddy-faced with a pursed mouth, sits cross-legged on several cushions, his knees pointing skywards. He speaks with a cultured accent. 'I've come to learn to meditate. I do have some stress in my life.' His body language indicates this is an understatement.

Queenie tells us, 'I'm a three-weeker and a runner.' I understand what she means: someone who doesn't see through the detoxification period. Three weeks is the longest she's ever abstained from alcohol, she tells us.

'I'm Shas. I don't know why I'm fucking here. Well, I do, I came 'cause I'm a friend of Cindy's. She said I needed to come.' Cindy is an earnest mother of two whom I met briefly last night, when she drove Shas here. Shas is small, in her twenties, with eyes like a possum's. She was an apprentice jockey when she fell off her horse at training one morning. 'I still don't remember what happened. That was eight years ago. Now I see things differently,' she says, but does not elaborate. She has a husband and a young daughter, and they are setting up a cattle farm on a bush block outside of town.

The previous night, after dinner, I came across Shas in the downstairs walkway, reclining on a couch. 'How ya goin', mate?' she asked, with a blunt look. She wore a weathered Akubra, a work shirt, shorts, and no shoes. The calf of one leg was heavily bandaged. I asked her what happened. 'Fell off the motorbike I got for Christmas, didn't I,' she says. 'Landed on the hot muffler. Not s'posed to be ridin'.' She sounded pleased with herself, and showed me the burn on her leg. It looked deep and nasty.

I doubt I'll have much to do with her, I thought. But later Choeying told me that she was the other brain-injured person attending the course. 'She needs a break from the family,' Choeying said, 'more than she realises.'

Shas was keen to talk to me when she found out that I also had a brain injury. Just this morning, she took me outside after breakfast and pointed up to the clouds, asking, 'What do you see?'

'I see clouds,' I said.

'I see bubbles,' she said, 'and they're moving. Sometimes, I don't know if others see what I see. Walls can dissolve; they don't seem solid. It's like I could walk right through 'em.'

Her neuropsychiatrist told her that the things she sees are not hallucinations, exactly, and she will have to learn to live with them. I asked Shas about her brain injury. She had been in a coma, and what she said suggested that the injury was worse than mine, involving more frontal-lobe damage. She'd have difficulty with planning, staying on track, and social appropriateness, I thought. 'I had medication for the hallucinations before,' she said, 'but I don't like what it does. It makes me dull. I can't feel. You know, my family says, "Why don't you snap out of it?" Ted — he's my husband — he gets frustrated with me all the time. I sleep a lot. That's why it's good to meet you. You understand, dontcha?'

'Yes, I think I do,' I said.

'You're not one of these bloody Buddhists, are you?' she asked.

'No, but I like some of their ideas, and I think they're good people.'

I recognised her experience: others frustrated with you, shut down by fatigue, out-of-whack with the rest of the world. 'When I had my stroke,' I said, 'it was like I was only in the present; I wasn't thinking about the past or the future. Everything seemed shinier, more alive, more interesting. And I wasn't afraid. But afterwards, I thought I'd gone crazy.'

Thinking about what Shas said, my mind has wandered from the introductions. Back in the meditation hall, we move on from Shas to a young woman who is an emergency-theatre nurse. She tells us that she loves her job but would like to cope with the stresses of her work better. Another woman is grieving over the death of her young son, two years before. She still feels her son's presence at home. The next woman has been diagnosed with cancer and fears the worst. She has an eighteen-month-old daughter.

A husband and wife tell us that there is a 90 per cent chance she has multiple sclerosis. They have desperately wanted a child but couldn't, so the likely diagnosis brings with it some bittersweet relief: no pressure to keep trying anymore. A woman with a closed expression tells us that she suffers from depression. Her brother was recently confined to a wheelchair. She has a baby and a toddler. A young man who sits at the front gives only his name: Josh. Choeying's already told me that he is one of her most committed students.

Tears have flowed in the room by the time it's my turn. 'I've had a stroke. My marriage is difficult. We're facing financial ruin. I'm really exhausted …' I say. Soon I'm crying too, but not from despair. It's relief — relief to be in a place where it is okay to let go and to be held in the supportive silence of others who know what pain is.

'I had lots of material things. The perfect life, perfect children, a beautiful home,' Choeying says. 'I was always helping others. I didn't need help myself. I thought these things would bring me true happiness. But my happiness was dependent: dependent, for example, upon whether I was meeting a friend for coffee that day, and if they didn't show up I was devastated. Even the coffee being bad could make me miserable. How can you live your life like that?

'We have to look at our minds. Our senses create desire in us. We like chocolate, but we can't just have one piece because desire takes over and then we guzzle and feel sick and wish we hadn't. But the next time we have chocolate, we have too much again. We like a particular type of music, so we listen to it over and over until it becomes dull. What we look at, feel, taste, touch, and smell is neither good nor bad. If we perceive it as good, we can't get enough. If it's bad, we can't get away fast enough. On the other hand, if we see everything as "just is", we see a new reality.

'I see, I like, I must have. The mind is not clear when it is driven by the senses. Meditation refines the mind, helping us to see clearly.'

She gives us the first set of meditation instructions.

DURING THE AFTERNOON session, I feel a tap on my knee. It's Shas — she's sitting behind me. 'I'm afraid,' she whispers in a quiet voice, bringing her face close to mine. She looks terrified.

After our talk this morning, I'm feeling protective of her. 'What's happening?' I murmur. Choeying is talking to the group.

'I'm not sure if I'm here.'

I reassure her that she's here and raise my arm to get Choeying's attention.

When I tell Choeying what is happening, she asks Shas to come and sit with her at the front. Shas settles on a meditation cushion beside her chair and, after a time, beams at us like a contented cat.

When the day finishes at three o'clock, I feel raw and exposed. The emotional impact of my revelations and those of the others is more than I had been expecting. I go for a long walk on the beach, where the gentle waves seem tender, and eventually I feel soothed again.

That evening, as I sit around the dining table with Shas, Queenie, Choeying, Rex, and Josh, it feels as if I'm part of a fraternity. We've all pitched in — helped to cook, set the table, and wash up — and now we're just chatting.

We're talking about what causes feelings. Choeying thinks that there is always a thought before a feeling. I say that neuroscientists cannot yet explain what a thought is, but they do say that emotions can come first, before a thought. I explain my understanding of the limbic system and how the nervous system enervates the whole body, including the organs, and these give rise to emotions. Choeying believes there must be something before an emotion and asks what is the origin of the limbic system. I suggest DNA and our parents.

'And before this?' she asks.

'Well, all the way to the beginning of the universe,' I reply.

But something must come first, she insists.

I have no reply. She's not convinced by my point of view.

Before bedtime, I Skype in Rex's office with Anna and the kids. Ashley tells me that she's made a video to the soundtrack of 'Sweet Dreams' by the Eurythmics. Anna says it's very good. Emma is already in bed, but Amelia is chatty. She is amazed that there was no sand at the southern end of the beach today. 'It was like a swimming pool, Daddy,' she says.

I am warmed and comforted by their conversation. It feels right that I take this time out to get better; I'm doing it for them as much as for me.

After we hang up, I retire to my caravan and turn in. I think I'll sleep soundly after the intensity of today. But I'm troubled by the discussion about the source of emotions. During the night, I get up and read the book I've brought with me — *Buddha's Brain: the practical neuroscience of happiness, love, and wisdom,* by Rick Hanson, a neuropsychologist, and Richard Mendius, a neurologist. I look for a reference to the neurological source of emotions, but instead get stuck on reading about the self.

Hanson and Mendius suggest that the concept of the self is an aggregation of brain states. A person begins, as a baby, without a sense of self. He or she forms one over time, believing that there is continuity in brain states, rather than each existing only in infinitesimally small slices of time. In other words, the self is a concept we manufacture, an attribution we give to our brain states, they say. The Dalai Lama said something like this too, although he did not put it in neuroscientific terms. Interesting.

I'VE NOTICED SINCE I've been here that Choeying's students pop in at any time of day. She doesn't seem flustered by it, and takes time to speak with each of them. She's clearly a valued member of the community.

The next evening, one of the younger students in the course, Alena, joins us for dinner and brings her steel-string guitar. The top string has broken: it has only five strings instead of the usual six. She sings two of her own songs in a soulful and tuneful voice. I take a turn and sing 'Father and Son' by Cat Stevens. Alena begins to cry and says how beautiful the words are. Josh, a classical guitarist, tries playing a piece that would be challenging enough on a steel-string guitar, let alone one with only five strings. Then we switch to golden oldies and folk songs, and everyone sings along. Rex sets up saucepans of water on the table, which he bangs with a wooden spoon. Choeying taps metal spoons together energetically. Shas jiggles a rice-filled jar, and Queenie sings along in a soprano voice, surprised at how many popular songs she knows, given her Baptist upbringing.

I look at the faces around the table and see such happiness. I have that 'in the music' feeling, where ego disappears. I'm struck acutely by the thought that making music is really about bringing joy to others; it's not about being impressive, or even about self-improvement.

The next morning, I'm mentally fatigued but happy. Choeying tells me, 'I saw the real David. I saw someone who was really enjoying himself. When you're singing, there are bigger moments of seeing you. Your eyes twinkle and you look so young. This is why chanting is valuable, too.'

I'VE MADE A decision not to watch or read any news while I'm here. I want to rest my mind and keep mental disturbances to a minimum. I hope that this will intensify mindfulness: I want to make the most of my time here.

We've had relatively little rain in Hervey Bay since I arrived. Rex explains that this is because Fraser Island, just off the coast, provides a buffer to the weather. But it's raining all around us, he says, and floods are causing widespread devastation throughout south-east Queensland.

That morning, Choeying has those of us at the house chant with her in the meditation hall. The chant emphasises the recognition of suffering in all living beings; it's a means of cultivating compassion in us. We focus our minds especially on the flood victims. Afterwards, I'm so knocked out that I take a longer-than-usual midday nap in the caravan.

That night, there is lots of talking and laughter among Choeying, Queenie, Shas, Josh, and Alena, who are sitting on the upstairs balcony until late. I'm trying to sleep again. The humidity, the mosquitoes, and the lack of breeze make it difficult anyway, without their noise.

As I resign myself to being awake, my thoughts begin to drift. I wonder, not for the first time, what sort of Buddhist nun Choeying is. She sees an endless stream of people during the day, stays up late, laughs lustily, cracks jokes, and starts the routine all over again the next morning. She doesn't seem to run out of energy. She always allows each person her full attention and is unsparing when giving advice — some end up in tears. She has already told me, 'You tend to be the teacher rather than the student in every situation. Until we can be the student, the one that's not coping, we won't get better.'

I can see how she would've been a charming and savvy businesswoman.

I have asked her about this capacity to be with people, to listen intently to their troubles and not be sucked dry by them. I remembered how worn out I'd be, when I was working in my practice, after only six hours a day of hearing others' stories.

She told me that she feels endless compassion for others' suffering, and that she finds this uplifting rather than draining. 'We can't have compassion for others until we know what it feels like to be rejected, criticised, abused, and judged. If we don't, and someone shares their pain with us, we'll say: "Just get over it."'

I also asked her how she handled the distress of inmates at the prison when she was visiting. She said that it required empathy, but you needed to move through empathy to compassion. This is 'empathy with the view'. 'The view' is the Buddhist way of understanding existence and human suffering. We are ultimately responsible for ending our suffering. 'Empathy is the fertile ground to grow compassion. You had the empathy without the wisdom,' she said.

I've clearly got a long way to go before I have the type of compassion Choeying has.

SEVERAL DAYS INTO my stay, I have a one-to-one session with Choeying in the meditation hall. I've asked if we can review my meditation practice, and I have some specific questions.

Choeying suggests that, in addition to the calm-abiding meditation, one could extend the object of meditation to every sensory experience simultaneously. This is difficult, she says, but it trains the mind.

'I've been thinking about what you've been saying during the course about what happens with meditation,' I say. 'I think I experienced these things when I had the stroke.'

I describe what it was like being in the waiting room at the hospital and during the investigations with the doctors: how

I wasn't perturbed by anything (except for the noise of the breathing machine from the patient in the bed next to mine). How everything and everyone seemed fascinating, and I didn't have a sense of people being good or bad, inferior or superior, and wasn't troubled by thoughts of whether I liked them or not. Time became irrelevant. I felt present in each moment as it unfolded, thinking neither about the past nor about the future.

'It's remarkable. Your experience … meditators take years to achieve these things. I've been sitting on my bum for years, looking at my crazy mind to experience something as profound as you have. Wow. A knock on the head: it's a fast track to enlightenment!' Choeying says. We both laugh.

It sounds preposterous, but I think there's some truth to it. I've only seen my stroke as something bad that has happened to me, as something gone wrong, but now I can see it as something special. I decide that while I'm here I will write an account of my experiences immediately following the stroke.

After my session with Choeying, I go for a swim at the local pool. As I do laps, my mind hums with reflections, my body enlivened. I have experienced the intense mindfulness that long-term meditators aim for, a taste of nirvana, and it's real — it's achievable. I've done it once, and perhaps I can do it again.

17

IT'S A NEW day. As part of my morning meditation, I've been doing a loving-kindness practice. In the hall, I sit cross-legged and begin. I visualise someone that it's easy to have an unconditional feeling of warmth for — my father — and then draw the feeling of warmth and loving-kindness into myself. It feels awkward, as if I don't deserve these emotions, but it's getting easier.

Then I move on to someone close to me (Amelia), projecting the feeling of warmth onto her. Next, I think of someone neutral (the waitress in the cafe yesterday), and, lastly, someone I find difficult (the creditor I had the barney with over the telephone), projecting feelings of loving-kindness towards them both. I have been including Anna, the children, extended family members, close friends, acquaintances, and strangers in the practice.

As I finish, I remind myself that I'm doing everything I can, and that it's okay to feel inadequate and lost at times. More and more often, I'm believing this.

Then I reflect on my progress in recent months. My brain is behaving better. Before I left, I completed the Brain Fitness course, reaching the higher levels of achievement on all six exercises. ('Tell Us Apart' — the exercise that required me to distinguish between similar-sounding phonemes — remained the toughest, and my weakest.) Doctor Small said there would be natural improvement with time, and I'm sure I'm seeing the benefit of this, too.

I haven't had suicidal thoughts for months. I still get in a flat mood sometimes and it can be an effort to do things, but I've worked out that this is mostly due to physical exhaustion and mental fatigue.

It is now eighteen months post-stroke. After the emotional storms of recent years, there are finally patches of blue sky. Following my talk with Anna last November, I realised that I'd been so caught up with survival — with simply getting through each day, with trying to meet the family's immediate needs, my rehabilitation obligations, and the legal and financial crises — that I had had little mental space for anything else. Lately, my mental space has expanded to include others; I want to know how they're going. These whiffs of expansiveness are like smoke from a fire I can't see, so I'm not altogether sure the fire is there, but it feels close by.

Before I left for the course, I was keen to check in with my daughters. As I was driving to a swimming squad with Ashley a few weeks ago, she'd said, 'Why do you get angry so easily?'

'What do you mean, "so easily"?'

'Compared to everybody else.'

Ashley doesn't often reveal her private thoughts, so I'd seized this opportunity. 'Well, I got traumatised from hearing about the bad things that happen to people — the people I helped at work. This made me irritable, you know, getting upset. Then I had the stroke and my brain wasn't working properly.'

'You don't even understand my question,' she said, and dropped into silence.

The door had closed.

Two months before this, we had been driving to her water-polo session in the late afternoon and the sun was in our eyes. 'That sun is annoying,' she said.

'How can the sun be annoying?' I said. 'It gives life, sustains us.'

'Well, you gave me life and you're annoying … Maybe not all the time.'

I laughed, and caught the flicker of a smile on her face. I like her quick wit. But I was disheartened by her comment. How has my irritability, fatigue, and self-absorption affected the family?

Late last year, I booked, after doing lots of research to find a place on our limited budget, a nine-day stay at a family resort in Fiji for late January. It was cheaper than a similar kind of holiday in Australia. I should be reinvigorated by then, after this break from the family. I want us all to have a restful holiday: relaxing in a natural environment, all meals provided, and no telephones, television, or computers. It will be an ideal way for the family to reconnect, and to assess what the damage has been.

Today, after a swim, I head into the police station. Due to the floods, I'm becoming concerned about getting home. The police officer tells me that the roads will be closed for at least three to four more days, and then they'll be in poor condition. He doesn't know when I'll be able to travel south. Nothing's getting in or out of town, and fresh food supplies are running short in the shops. There are signs of panic buying.

Upon hearing this news, I get that trapped feeling I had in Seaview. How will I make it back in time for the holiday? When will I get home? I also need to see my GP to complete the monthly claim form for the insurer so that I am paid. A delay could mean more financial disaster.

Rex suggests that I book a flight to Brisbane. The earliest flight I can get means that I will arrive a day late in Fiji. The family can go on before me. I speak with my claims manager, and he says that he doesn't expect a claim form this month because of the floods; others are in the same position.

The other out-of-town participants in the course are also 'trapped'. Choeying says that it's a wonderful opportunity for us all to stay longer; our employers will understand. It means I will be here for almost two weeks.

Choeying's right. With the assurance that I can join the family in Fiji, I feel better. And the almost daily revelations I'm experiencing are too valuable to miss.

Something Choeying said on the first day has stuck in my mind: a bad cup of coffee at a cafe would have once put her in a terrible mood. She's also given us homework: to notice how our feelings change in different situations and in response to different people. This is to encourage mindfulness.

My usual practice when staying in a new place is to ask a local where the best coffee is. If I like it, I keep going back there. I only allow myself one cup of espresso per day, so a bad one is disappointing. So I thought, what if, rather than re-create the safe experience of going to the same cafe with the best coffee, I go to a new cafe each day, and watch my reaction to the quality of the coffee, the service, and the ambience?

This has been my project since my arrival. The first cafe I went to, on the Esplanade, had good coffee. The service was fine, but the place was noisier and more cramped than I'd like. I gave it eight out of ten, and I would've been happy enough to make it my preferred cafe. But, staying true to my resolution, that afternoon I arrive at a new cafe. Immediately, I don't like the ambience. It has uninviting tiled floors, hard metal chairs, and no softness or quirkiness in the décor. But I push myself to go inside.

The older man serving me seems like the proprietor — he doesn't look like a skilled barista — and his lack of passion for making coffee is obvious. I can't stand to watch him, so I go and sit outside.

My first taste of the coffee tells me that he has burnt the milk, and the flavour is thin. Disappointment. I register my impulse to leave. But I breathe, lean back, and sit with the feeling. I look across the road and notice the greenery and the children's playground in the park opposite. I get a whiff of the sea air. People are on holiday, cheerful. The disappointment and the urge to leave fizzle out.

I sit for half an hour, and end up leaving most of the coffee. But I've enjoyed my coffee experience, and I've enjoyed watching my mind and investigating how tricky it is.

When the course participants meet for our mid-week evening catch-up, I say, 'I've been doing coffee meditations.' The others laugh. I explain my rationale and what I've discovered. 'I sit with the bad coffee and notice the initial disappointment. If I relax into the experience and I don't push the disappointment, or whatever, away, it fades. I can still enjoy being in the cafe. Although I'd prefer a good coffee.'

Each time I meet them after this, I'll be asked: 'David, how are your coffee meditations going?' They want to know my latest ratings.

IN MY PRIVATE morning meditations, I've been slipping into deeper and deeper states of mental stillness; the realisation that I experienced intense mindfulness during the first hours of my stroke has super-charged my meditations. I can sit for up to seventy-five minutes before it becomes physically uncomfortable. Often, I lose sensation of my lower body, or it feels distant.

To begin with, Choeying has given us a 'love and equanimity' meditation from Buddhist monk Thich Nhat Hanh, which focuses on ourselves. I understand this to mean that until I have true compassion for myself, I can't have it for anyone else. After we've focused on ourselves, Choeying says, we can focus on others. It goes like this:

May I be peaceful, happy, and light in body and spirit.
May I be safe and free from injury.
May I be free from anger, fear, and anxiety.
May I learn to look at myself with the eyes of understanding and love.
May I be able to recognise and touch the seeds of joy and happiness in me.
May I learn to identify and see the sources of anger, craving, and delusion in me.
May I know how to nourish the seeds of joy in myself every day.
May I be able to live fresh, solid, and free.
May I be free from attachment and aversion, but not indifferent.

I've been doing this, both for myself and for the family. It has a soothing, softening effect and, somehow, it warms my heart.

One morning after meditation, Choeying and the others decide to go into town to shop and have coffee. When I say I'd rather go out by myself, she suggests that I do a 'driving meditation' as another way of practising mindfulness. She tells me to cultivate the attitude that I'm not driving the car, the car is driving me, and to notice everything. 'And careful,' she says, 'you might get hooked.'

I sit in my car with her instructions in mind. Then I exit the driveway and turn into the road. I'm paying careful attention to the surroundings, in that non-judgemental way I experienced after the stroke.

By the time I'm on the main road into town, I'm in a tunnel, seeing only the road in front of me and the trees on either side. I realise how distracted I usually am when driving.

Soon I become completely lost in the process of driving. I begin to feel as if the car is an extension of me. I think *left* or *right* and it turns by itself. I know that my arms are doing the turning, but it doesn't feel this way. The movement is velvety smooth, as if I'm in a luxury vehicle, not an old Subaru. I am progressing through space on a cushion of air, the car a metal wrapper. My seat supports me lightly.

In this mental state, each rise and dip in the road becomes exhilarating, as if I'm on a rollercoaster, but without the stomach churn. When the car pulls up at the traffic lights by itself, I wait — with no impatience, content to be exactly where I am. Clock time is passing, but my time isn't: time has become irrelevant.

I reach the town centre, park, and pick up my flight ticket.

Then I go for coffee in a nearby cafe. For the first time since I've been in Hervey Bay, I pick up a newspaper; I'm intrigued to know how I will respond to it in this state of mind. Rustling through the pages, I read about the floods. I feel compassion for the flood victims, and try to contemplate what they're going through, but I'm not dragged down by my concern.

Emboldened by this, I turn to the business pages: a section I've read assiduously before for commentary on the housing market and what this will mean for our finances. An article says that interest rates may go up again. This will make paying off our loans even harder. As if from afar, I watch a disturbance gathering in my mind, like wind causing ripples on water. But the disturbance is still far off, and I put down the newspaper. With this and my coffee meditations, I'm impressed with how I can watch my thoughts and feelings come and go.

Finally I think I get it: mindfulness is not like an umbrella, which you only put up when it's raining; it's something that you have on all the time, as I did during my early stroke experience.

When I drive back, it is harder to get into the same mental

state as before. Perhaps the caffeine has kicked in, or reading the newspaper has disturbed my composure. I dip in and out of mindfulness, and purposely extend the trip to enjoy it longer. Choeying was right; I could get hooked on this driving meditation.

ON THE SECOND weekend of the course — the weekend before I leave for Fiji — Choeying has us focus on breath meditation. There is a lot of feedback from the other participants, but I'm content not to say much. During the lunch break I take my plate of food and sit on one of the chairs under the walkway. I look into the garden — the outlines of the leaves and stems are brilliantly clear, the colours rich. It's that same sparkle I experienced in the hospital after the stroke. Again, time seems irrelevant; I'm simply content to *be*.

I'd said earlier to Choeying that the other participants were not people I'd usually spend time with or come across in my social circle, but I felt close to them. My sense of companionship with Shas has shocked me; we have very little in common, yet we've hung out a lot together. A few nights ago, Shas invited my family and me to come stay on her property. 'We can go chain-sawin',' she says. I think this means we're mates.

I've noticed that the hardness in her face has gone, and her conversation is more considered.

Choeying had said, 'We hang out with people the same as us; we limit ourselves, and that's why we don't grow. You are realising the benefits of equanimity, darling.'

At the end of the day, my last on the course, there are hugs. I hug the woman with probable multiple sclerosis and the woman with cancer. I feel for them, but I'm not dragged down to a place I don't want to go in knowing what they're facing. And it seems as though I'm giving them something through my presence, though I'm not sure what that could be.

On my last night, I read aloud my written account of the first twenty-four hours of my stroke at the dinner table. It causes some laughter. Choeying is moved, and Shas says she's grateful to have heard it. I talk about Jill Bolte Taylor's account of her stroke, and how she'd had some kind of remarkable experience — but I couldn't remember what it was.

After dinner, I do an online search and come across Taylor's TED talk. It's mind-blowing. She says that the serial-processing style of the left cerebral hemisphere leads to a strong sense of the individual self, distinct from others. The left hemisphere concerns itself with the past and the future, and how the present moment relates to these. The right cerebral hemisphere, on the other hand, with its parallel-processing style, has a strong sensory, in-the-moment way of experiencing the world. On the morning of her stroke, Taylor suffered a left-brain haemorrhage, which led to the silencing of her internal chatter and the (temporary) loss of thirty-seven years of 'emotional baggage'. She felt extreme peacefulness, enormous expansiveness, and compassion for all beings; and, in between these states, periods of panic as her left hemisphere kicked in and she tried to call for help (she was living on her own). Yet she describes her right-hemisphere experience as like discovering nirvana, and is overcome with emotion in speaking about it. After what I went through, I know that she's not making it up; her experience contains many of the elements of my own.

The evening before, I'd suggested that Choeying, Rex, Queenie, Shas, and I go for a barbecue by the bay. During my walks I'd discovered a picnic table overlooking the water, with a covered barbecuing area nearby. A middle-aged German couple, Arnold and Karla, had joined us. They hadn't been on the course, but I'd met Karla at the house already.

Rex and I cooked in the BBQ area. There was a Christian group sitting in the adjacent shelter, and I heard them talking, over glasses of wine, of their experiences on some or other recent camp.

Our group sat at the picnic table by the wall of rocks overlooking the beach. It was dusk, and an orange light suffused the sky over the western side of the bay, which was hanging before us like a canvas. Along the curve of the beach, in the distance, there were two fishermen, barely visible against the greying water, their rods thin as whispers. Choeying and Karla sat on the grass facing the beach — profiled by the fading light, their legs dangling over the wall as they chatted, bent towards each other like old friends. Arnold stood by the table, tall and lean, speaking hesitantly and picking at the food. Queenie and Shas sat side-by-side, tucking into a bag of potato chips, just like kids; the branch of a she-oak arched over them like a wispy parasol. Queenie said that it reminded her of Friday nights, when she would start drinking and eat 'chippies', but tonight she felt strong enough to resist alcohol.

As we talked, a three-quarter moon rose over our shoulders. The fruit bats chortled like goblins in the trees above. *This is my family,* I thought. I delighted in their chatter. There was warmth in my heart and peace in my mind.

I was different from the person who, beached by exhaustion, feeling pulled apart, had turned up at Choeying and Rex's doorstep a fortnight before. A previously unopened door had been opened; a new mechanism had been inserted in my brain. Without needing to analyse why, I knew that I was shedding an old cocoon.

I have learnt firsthand that the mind is like a palette of colours; two people can look at the same scene and one of them fills it in with bright shades, while the other draws upon sapped-out tones and hues. The externals have not changed, but the view for me, looking from the inside out, has transformed.

18

THE DAY OF the dive is as humid and lacking in breeze as most of our days at the Plantation Island Resort in Fiji have been. Silver air tanks and black buoyancy vests stand to attention along the centre line of the boat. Snorkels, masks, and fins sit at our feet. The black Yamaha outboard rouses with a throaty roar, which turns to a gurgle and a splutter once lowered into the water. We edge out from the jetty, petrol fumes washing over our faces.

I am sitting on the bulwark of the boat, opposite Ashley. A tired canopy shades us. Two others are diving with us: a young man from New Zealand who works in Fiji, and the dive master, a Fijian with an afro and shoulders like a rugby player who doesn't speak much.

Once we round the jetty, we thread through the families laughing and splashing in kayaks, and make it into deep water. The bow lifts with the pressure of the water beneath as we increase speed. Salt spray cools our hot skin.

The breeze pushes through Ashley's wavy hair and against her face. She looks at home.

I am pleased. I knew that she liked snorkelling. Once we'd settled in at the resort — me arriving a day late — I'd suggested that she take the Open Water Diver certificate course, now that she was old enough. Over the last five days, she attended a class in the morning and then practised her water drills or dived after lunch, coming back each time telling us how good it was.

Today's dive is her last to complete the course. It is the second-last day of our holiday. I had asked the instructor if I could join them; I wanted to share this experience with her. I have my certificate, but it's been a long time since my last dive, and I was apprehensive this morning. Still, I ran through the safety checks with the instructor and he allowed me out.

Ashley and I are going to be dive buddies. We will be entering the water around a dot of an island, one completely ringed by a coral fringe: an easy dive.

A few days ago, I said to Anna that I'd like to talk with the girls alone. The four of us had taken a mid-morning walk to the drink stand by the beach, where we ordered cold drinks. Even the air felt lazy. We sat down at a plastic table, lodged in the sand and shaded by a wide-hanging eave.

A young couple sauntered by.

I felt nervous about this conversation: how to start it and how to encourage them to talk. To ease my nerves, I tried to be fully present with them, as if nothing else existed, rather than trying to second-guess what they might say and how I should respond. It was another of those little moments of standing on the edge and having faith in the process, I told myself. Since Choeying's, my mental space had expanded well beyond me: a circle of embrace I had not felt in a long time. As we sucked on our drinks, I had the strongest sensation that I was breathing with them, as if our physical boundaries had merged. I could take in anything they might say — confronting, unpleasant, or otherwise.

'Girls, I know I have been grumpy and shouty lately. I was very tired last year, before Christmas. I had a good rest when I went away. I'm much better now.' They look at me. 'I'd like to know what it's been like for you when I get grumpy and shouty.'

'I got scared when you had your tantrum in the kitchen,' Emma said. 'I thought you might hurt us.'

'Amelia, is that how you felt too?'

'A little bit,' said Amelia.

'And Ashley?'

'I don't think you love us,' Ashley said.

'I only thought that when you were angry with me,' said Emma.

I told them I would never hurt them, and when I was angry, I was mostly angry with myself. They may have done something that triggered my anger, but they were not to blame.

'This is boring,' Amelia said, clunking her head dramatically onto the table, rolling it from side to side in mock agony. 'Yes, we get it. Can we talk about something else?'

I suppressed a smile. I wasn't finished. I needed them to understand; otherwise, I would tip over the edge again and there would be another 'tantrum'. 'When Mum or I ask you to do something, we need you to do it the first time, not the second or third time, otherwise I get exhausted and end up shouting. We need you to do more jobs at home; I can't do as much as I used to. I love you all *very* much. I think you will turn out to be very fine people.'

'You can't predict the future,' Emma said.

'I *believe* you will turn out to be fine people,' I said.

I felt easier after this conversation, and as far as I could tell, they did too. But I've learnt with my kids that their actions often reveal more than their words, so time would tell. The best I can offer them is to be as present with them, in the mindful way I've learnt, so that they feel listened to and understood. And with my newly behaving brain, I'm confident I'll be more even-tempered from now on. I am regaining control. We are turning the corner.

The water is like turquoise glass, suspending the boat. The horizon shimmers in the humidity. After forty minutes, we reach the island — a tuft of palm trees and starch-white sand leaping out of the water in the middle of nowhere. The dive master explains that we will dive down to our full depth and then drift with the current along the coral shelf as it circles the island, slowly rising to meet the boat.

It all sounds logical and reasonable, but once underwater I am apprehensive again. I'm conscious of the volume of water above us and the heavy lump of tank on my back. The regulator is uncomfortable in my mouth: each intake of air is harsh and rasping, the air bubbles released loud in my ears.

Ashley is nearby; she looks at ease.

Anxiety, I remember, escalates the use of oxygen when diving. I turn my attention to the sensations of my breathing and slow the rhythmic beat of my legs. It works. Soon I am able to take in the splashes of colour from the coral formations, which are almost preposterous in their diversity of shapes and sizes. Schools of fish parade before me in curtains of iridescence, as if cavorting in a fashion parade.

I'm hovering over a large fan coral, fawn-coloured with blue edges, when a hand slips into my left: it is Ashley. I give her the diver's okay sign with my right hand, and she returns it.

Together, like this, the gentle pulsing of our fins propels us along. A large parrot fish stalls in front of us, a puff of excrement exploding from beneath its tail before it moves off. Ashley and I turn to each other, giggling in our masks at the affront. I feel a thrum of love flowing through the conduit of our joined hands.

The dive master signals for us to surface after only half an hour; I still have almost half a tank of air left. Back on the boat, the New Zealand man reveals that he ran out of air, initiating the order to surface. I thought I'd be the most likely to have this problem. *Wow*, I think, *I must have been more relaxed than I suspected.*

THE NEXT DAY, we ride the catamaran back to the mainland. It has been a wonderful holiday. Not only do I feel a new ease with the children, but also a softness towards Anna. We'd shared the same bed, socialised with others together, and taken walks with the children. Our conversations had become easier as the holiday went on.

Anna enjoyed the socialising and the water sports. Emma and Amelia liked the swimming pools, the water sports, and the painting of T-shirts in the kids club. Ashley liked everything and wants to live in Fiji. We all liked Maurice, the evidently gay waiter, who wore a fresh hibiscus flower behind his ear, each evening offering us the same drinks menu with a flourish and talking us through the limited dinner options (we were on the budget-meals package) as if it were fine dining.

We had played cards in the late afternoons at the tables set up on the beach, the rising tide surrounding us in ankle-deep water, the stunning sunset our backdrop. Guests had taken photographs of us. I had even played with the four-piece band for two nights, when the main guitarist was on leave. The good-humoured dinner audience had gotten a big dose of James Taylor, and the boys in the band had enjoyed learning some new songs.

And we'd had several experiences — the type of family memories that grow more amusing with each retelling. One of these had taken place when we attended a Sunday church service promoted at the resort, thinking we'd hear wonderful islander singing.

The small, picture-perfect church sat on a knoll — the highest point on the island — catching the languid morning breeze.

The pastor, a very square, unsmiling man, wore a black jacket over a white-collared shirt and a plain sulu. For much of the service he stood at a high, imposing pulpit. Only a few of the more elderly locals attended, outnumbered by us tourists. (We learnt later that many of the locals attend an earlier service, before work.) As a consequence, the singing of hymns was on the thin side, carried by the few tuneful regulars. The second half of the hour-long service was given over to the pastor's sermon — in Fijian. Like a slow-building storm, his demeanour and voice took on an increasing menace as his words rolled on. It reached a crescendo with him shouting, in eye-popping fashion, at the congregation, apparently castigating us for sins that we, at any rate, couldn't comprehend.

I turned to look at the girls seated in the pew behind Anna and me. Emma had stuffed the beaded ends of her recently braided hair in her ears. Amelia's hands were clapped solidly over her ears, her elbows pointing forwards, while Ashley pondered the floor with great interest, her hands twitching in her lap. They all grinned when I caught their eye. I couldn't know what the pastor made of the sight of their disinterest, but I suspect it only increased his fervour.

'That was boring,' Amelia said after the service, when we'd emerged into the sunlight to mingle with the other rattled-looking visitors.

'Yes,' I agreed, chuckling. 'But we've had a new cultural experience, and that's something.'

WHEN I CATCH up with Doctor Franklin a few weeks later, in February 2011, I have much to tell him. I meet with Doctor Franklin every month — he calls it a psychiatric review. He has become a helpful advocate. I alternate his session with Wayne's so that I see one of them every few weeks, although I speak with Wayne more often than this if something comes up.

Doctor Franklin's consulting style is pragmatic, flavoured by his years of experience. He doesn't profess to be a psychotherapist and he encourages my contact with Wayne. He visits a country town, less than an hour's drive from my place, one day a week, and it is here that I go to see him. He's become one of the doctors, together with Doctor Small and my GP, who can fill out the treatment report for my monthly insurance claims.

We sit alongside his desk, our chairs facing. He takes few handwritten notes, so we maintain eye contact most of the time. I tell him that I've written about my stroke and that writing is helping me to join the dots in my story, giving me new insights. In our financial dealings, the dread does not overtake me as much, or for as long, as it used to. I am more confident and assertive, less like the sea anemone that closes up with a hint of threat, and the stretches of confidence remind me of how I used to be. I haven't had any thoughts of escape or suicide since last year. I have been writing summaries of my readings for James, and he seems pleased with them.

Doctor Franklin says there are less bushfires in my life now. Each property, each financial threat was a bushfire. Dominique and her associates are helping us to fight the fires.

I am also less reactive. If Anna or the kids say or do something that winds me up, I am alert to the feelings arising in my body. Often, I can pull myself up before I say or do something impulsive. Sometimes I don't respond outwardly at all: I let it ride. Anna has told me, 'You're calmer since you came back from Choeying's.'

I tell him that I have renewed hope for Anna and me. With the kids, there's been a renaissance: they are more cooperative and loving at home. I feel what they're feeling more. Ashley hangs out with me now — I take her to singing lessons and accompany her on the guitar sometimes, and she often comes to swim squads with me.

For the first time, I ask Doctor Franklin how he approached the diagnosis of my condition when I was in the psychiatric clinic. He says that he first thought my symptoms could indicate something neurological in origin: a stroke, epilepsy, or something else. If not this, it could have been a case of malingering, or psychological dissociation (a fugue state). He needed to rule out the neurological possibilities before considering the other two.

It feels odd to hear that he considered, even if for only a moment, that I might have been malingering. But once again, I am heartened to have it confirmed that he did take a thorough approach to my diagnosis. How far back in my progress would I be if I'd continued to believe I'd had a mental breakdown and had never started on the Brain Fitness program? Life can be such a serendipitous journey at times.

BY THE TIME I see Wayne in April 2011, things have changed a bit. I tell him how the mental fatigue went away while I was at Choeying's and in Fiji, but it has returned to a degree since. Yet the girls are helping out more at home, so I'm not being pushed over the limit.

Wayne reminds me that each phase of the rest–activity cycle that goes on throughout the day and night is ninety to 120 minutes long. He recommends that when I'm working or involved in something requiring concentration, I rest every ninety to 120 minutes.

I have already worked out how to manage my mental fatigue more precisely, and it's really helping. I limit the work I do for James to nine hours, in three-hour blocks, per week. Otherwise, on top of everything else, it causes rubber brain.

Wayne also affirms the value of writing about my experiences. Psychotherapy is a search for meaning, he says, a way of finding a new narrative. I tell him, as I have Doctor Franklin, about my ideas of becoming a science writer, a neuropsychologist, or a neuroscientist. I'm investigating courses I might do to achieve some of these aims. 'When you mention writing, you light up,' Wayne says. 'But not when you talk about study or clinical work.'

Wayne's encouragement spurs me on with my writing, but it becomes harder. I write about my memories of the offenders I met in jail — this is the first time I've felt able to do it. It brings on light-headedness, a feeling of unsavouriness and distance from reality. But these emotions only last a few days, and I am able to move through them. I see this as a good sign. When I tell Wayne, he says it is an achievement that I can approach these memories and process them more fully.

Meanwhile, the other source of my anxiety, our financial affairs, won't go away. In May, I tell Doctor Franklin that I'm getting feelings of hopelessness again, triggered by our financial affairs. We are being sued, and chased again by creditors — there is only so much Dominique can do for us. And Anna is talking about separation once more.

In the July school holidays, we go on a camping trip to the Outback. It will probably be our last family holiday, as the kids are growing older and Ashley is showing less interest in doing things as a family. It's another attempt to get us away and into nature so that we can continue to reconnect. In particular, I'm keen to have time with Anna, so we can talk again. After Fiji, she had said that she wanted an eventual separation, and that she didn't think she could try again. However, she preferred us to keep working together as a team to deal with the financials and the children.

But I still don't want to give up. I'm hoping that the new me will make a difference.

Mindfulness, I'm finding, throws up everything: an awareness of good intentions, bad intentions, loving thoughts, angry thoughts. It seems to drill down into ever-deeper layers of mental experience, and it sharpens the relationship lens. I see clearly now, in a raw, sensate way, how emotionally cut off Anna has become from me: how she has been building a social life outside of ours, and how her business interests preoccupy her, and take her away as well. She doesn't seek my company anymore. Yet on the Outback trip it seems to me that Anna is like her old self.

One steamy morning, we shoo the kids away from our campsite and they go to cool off under the huge sprinklers set up in the camping ground, still within sight. 'Where is your head at with respect to us?' I ask.

She recites the familiar themes: there's 'too much history', we're 'a sexual mismatch', she has 'no desire' for me and 'can't see it coming back'. She wants us to be 'the best friends we can be'. She'd like us to begin living apart when we get home.

'I'm fifty–fifty about separating,' I say. 'I'm confused.' But even as I say this, I am reluctant to let go; I feel the tug of grief, now that it all seems real.

For the rest of the day, a deep sadness swamps me, dulling the landscape. I want to be alone, my mindset clearly out of sync with the kids' upbeat energy. Why is Anna so steadfast in her rejection of me, so impervious to everything I try?

Late that afternoon, we travel to a spectacular rock escarpment in Kakadu National Park to watch the famed sunset. There must be at least a hundred others gathered here as well. On a ledge below us, as the perfect ball of smouldering red pauses on the horizon, a young man gets down on one knee before a woman. There is the glint of a ring as he takes her hand. She accepts. Spontaneous applause and shouts of 'congratulations' and 'how romantic' rise from the crowd. Champagne is popped.

It's a bittersweet moment.

When we get home from our holiday, Anna sets up a separate bedroom downstairs. I don't know if the two are related, but this coincides with an increase in my mental fatigue, and a deterioration in my memory and my capacity to understand what I'm reading for James.

Then, out of the blue I get a call from Choeying. She is holding a get-together on a student's rural property over the weekend and invites me to come and join them. I go, and as before, it's uplifting to be with her and the other familiar faces. Talking about my grief and being with them eases the sharpness of my pain. Choeying and I also catch up with Shas, and she takes us for a drive on her bush property. It's not yet cleared, with cattle scattered among the trees: she and her husband have a lot of chain-sawin' to do, evidently. Shas tells me that she is sleeping a lot, getting fatigued again. I feel her confusion, and notice the strain on her face.

In my next session with Wayne, I describe what has been happening recently. He says that I could feel as if I've let the children down the more I go over what we've lost. I should remind myself that I'm still there for them. I'm alive. I love them. I'm a good man. I'm here and hanging in. If the girls see Anna and I each surviving this hard time and they still feel loved, they will be okay.

Wayne is worried about my cognitive deterioration. We both wonder: have I had another stroke, a small one? He suggests that I follow up with Doctor Small.

The next day, I do so. Doctor Small orders another brain MRI (my second) and the results come back: I haven't had another cerebrovascular accident, but the evidence of the old one is still there. Doctor Small says that the hole in the head will remain. So it's the impact of life events; my brain, hollowed in part though it is, is actually okay.

I give up on the plan to retrain as a neuropsychologist or a neuroscientist. It's become clear I couldn't manage the conceptual demands and the concentration for high-level study, or the long days. I begin using the word 'disability': a term I haven't thought applied to me before. It's one that others take notice of, even if they can't appreciate the exact nature of my disability. I find that when dealing with insurers, lawyers, government services, and some health practitioners, I need to wave this term in front of them, like a big stick, to get them to re-adjust their expectations of me.

The two things that remain most fraught for me are telephone conversations and being asked to make a decision on the spot. If it's something important, I ask the other party to put in writing what they want from me, or what they're trying to explain. I tell them I will respond in writing. This reduces the likelihood of me becoming overwhelmed. (My developing writing skills have become a real asset.)

It was Doctor Franklin who suggested this. 'If you try and justify your position, it means people can debate it,' he told me. 'Say, "I only deal in paper" and repeat it like a broken record. This is what I do. "Sorry, I can't help you further."' He's written a letter for me, which says: *In view of Mr Roland's range of difficulties, resolving issues should only be done in writing. I have advised him not to engage in telephone interviews.*

Over the last year, I've attended a few short writing workshops, and I've engaged a mentor through my local writers' centre. He has encouraged me to write in a stream-of-consciousness way about my life. I find myself writing about childhood times — lots of happy memories. It reminds me of how loved and safe I felt with my parents, and how, in their different ways, they contributed to my resilience and sense of curiosity; how proud they would be of me for not giving up, and for making something good out of the bad I have been through.

In August, I attend a session at a local writers' festival. Unexpectedly, one of the panel members, a journalist who writes novels, describes in graphic detail the punishment meted out to a nine-year-old boy by his mother and stepfather. She tells us that they suspected he had sexually interfered with his younger cousin. It later turned out that he had not, and they had based their conclusion on a throwaway remark made by the girl. Their extreme punishment led to the boy's death, shortly after he had been hospitalised. As I sit among the audience, the familiar attendants — disgust and horror — come. I contemplate walking out, which would bring attention to myself, so instead I remain, wondering if I can manage this type of talk now.

That night, my sleep is disturbed by bad dreams. I'm unsettled. I wasn't able to speak with the author after the session, so the next day I email her. I explain the effect her story had on me, and probably others, suggesting that she warn people beforehand if she is going to go into detail like this again. She responds quickly, apologising. She says she will take up my suggestion for the future, and praises the efforts of 'frontline people'.

The author's graphic account intrudes in my mind, on and off, for several days. But it doesn't derail me. When I talk with Wayne about the incident, he says, 'You're a helping professional. You've got a heart, you have empathy, and you feel the suffering of others. I would have had trouble staying there. We need to find a way to protect ourselves while keeping our hearts open to what others are going through. We have to accept the suffering out there.' He says that I recovered through my own efforts, regained my emotional equilibrium; this shows that I have made progress.

SEVERAL MONTHS BEFORE, I had attended a national psychotherapy conference in Sydney with Ian. One of the keynote speakers, Ajahn Brahm, a Buddhist teacher and the abbot of Bodhinyana Monastery in Western Australia, spoke about his approach to therapy. I liked his personable style: he told stories, cracked jokes, and explained his ideas clearly. I found out there that he was running a nine-day silent retreat near his monastery in late November. I've never attended such a long retreat before, or a silent one, but now I wonder if it could change me in some way that would make a difference to Anna — a last-ditch effort to save our marriage. Also, it would be good to have another extended rest. So I go.

On the third day, we have the opportunity for a short private interview with Ajahn Brahm. ('Ajahn' is an honorific given to a respected Buddhist teacher in the Thai tradition.) I am the first to meet with him. I confess that my marriage is in jeopardy; I haven't been as open-hearted with my wife in the past as I would've liked. She wants to separate, and I'm concerned for the children.

'Meditation will soften you up,' Ajahn Brahm says, and laughs. 'I don't know why husbands and wives are so keen to get my advice — I'm a celibate monk!' But, becoming more serious, he continues: 'Between couples, you don't want there to be any fear about saying anything. Tell your children that they can say anything to you or your wife, and although it may hurt at times, they will not be punished. Emphasise that it is Mum and Dad that need to work it out together.'

I mention that I'd asked Anna in a letter, which I'd left for her before I came on the retreat, to tell me the ways in which I've hurt her; I want to know what it's been like for her.

'It's not your fault, or her fault; it's nobody's fault. Be kinder on yourself. Her heart needs to be open as well,' he says.

I see that he's right.

I'm exhausted, and after this I go to bed and sleep for a long time. It amazes me how tired I am — I've slept a lot over the first three days already.

In my second private interview with Ajahn Brahm, I tell him how I've been through some terrible times in recent years and had to find a way out or I wouldn't have survived. 'I've learnt that pain is a strong motivator,' I say.

'Yes,' he says, 'but you need joy to stay on the path, to keep going.'

I recognise the truth in this immediately. The painful experiences motivated me, out of desperation, to find a way through the pain: to find calm, nurturance, and a way to insure myself from the outside attacks. I've experienced joy through singing, making music, swimming, movies, yoga, friends, family holidays, and, increasingly, meditation. Often, I can experience contentment despite outside events. Choeying's words, that suffering can be a 'blessing or gift', make sense now: pain is a great motivator to be a different person, to see things differently; either that, or you roast inside.

On the last day of the retreat, I take a tour through the monastery with some of the other participants. It is run in the Thai Forest Tradition, so monks live in their own small huts in the forest, with walking paths leading to the common areas, including the kitchen and the dining area.

Ajahn Brahm has mentioned that he sleeps in a cave; I wasn't sure if he was joking. We are guided by one of the resident monks, who speaks with an Irish accent. The cave is real, but human-made, and has been carved into the side of a small hill. A few toy plastic bats hang from the rocky ceiling. With the door closed, it is completely dark inside, unless the lights are on. The monk points out the thin mats that Ajahn Brahm lays upon the floor to sleep on.

We are shown into the small office space at the other end of the hut, where Ajahn Brahm writes his books and correspondence. On the shelf above his narrow desk are his publications and volumes of the Buddha's teachings. There is little else in the room, apart from a chair and a low bench seat. As we step out, I remark, 'There's not much in here, is there?'

'It's important to note what's *not* in there,' the monk replies.

This comment stays with me on the flight home. I feel a sudden keenness to simplify my life, to get rid of 'stuff': not only material possessions, but also the collected years of professional documents I no longer need; and, most importantly, frivolous activities or relationships that aren't nurturing and meaningful.

In his morning talks, Ajahn Brahm spoke of how not letting go can sustain pain, hurt, and anger, and reinforce worry. Now I am ready to let go of being a psychologist — both the one I was before and the variations of this I'd considered becoming.

And for the first time, I can let go of Anna without feeling as if I'm being sucked down a sinkhole. If she believes that she can achieve happiness by starting afresh, how can I say whether this is the right or wrong thing to do?

Is this surrender? Is it acceptance?

The holidays away where I thought we could be our old selves did not make a difference; they merely delayed the inevitable. But I have tried. My efforts to understand what she's been through have been thwarted, and I may never know what her experience of our marriage has been like. Yet I see now that I have been like the passenger standing at a bus stop waiting for a bus that's already gone, and looking expectantly up the road for it to lug around the corner. I will need to let go of the romantic ideal of the nuclear family that grows up together, sharing birthdays and Christmases, relatives and family friends. This is a sadness that will be difficult to sit with, but I've learnt a way through pain.

When I return home, Anna and I discuss how and when we will separate. I will see where this takes us. Let it unfold.

19

EARLIER IN 2011, James Bennett-Levy learnt that a highly regarded German neuroscientist, Tania Singer, would be holidaying in our area before Christmas. He made contact with her and pinned her down to spend a day with us. He's invited a few of his colleagues, and me, to come along.

We meet up at his home and are all quickly on a first-name basis. Coffee orders are taken, and I go out with someone else to pick them up from the nearby cafe. When I get back, James has set up a projector in the living room. Once Tania's computer is plugged in, the opening graphic of a PowerPoint presentation projects onto the pale-yellow wall, left bare after the removal of an Indian tapestry. It says: *Empathy and Compassion from the Lens of Social Neuroscience – Its Measurement, Modulation, and Plasticity. Tania Singer, Max Planck Institute for Human Cognitive and Brain Sciences, Leipzig, Germany.*

We settle into chairs, forming a rough semi-circle around Tania, while she stands. I like her immediately; she lacks pretension, and she smiles a lot. She is tall, fortyish, and speaks in German-accented English. Once a research psychologist, she is now the Director of Cognitive Neuroscience at the Max Planck Institute. She's refreshed from her holiday, which, she says, included a self-imposed ban from internet access. She's very pleased to have this time with us.

Her primary research area is affective neuroscience — the brain and emotion. Most research in cognitive neuroscience, she tells us, has been done on processes such as memory and attention, but there is little on 'affective networks', except for fear and psychological trauma. She's previously researched the human emotional response to pain. Today, she'd like to tell us of recent unpublished research on empathy and compassion.

In general, she says, it's more useful to think in terms of neural networks than fixed functions in localised parts of the brain. One area of the brain can serve different functions depending upon which network it is involved with. Her description fits in with my road analogy of the brain, with major and minor roads, crossroads and road junctions.

Emotional contagion, she says, is the precursor to empathy, and we share it with most mammals; we pick up on others' emotions unconsciously, and without realising the source of our emotional experience. If we are around other nervous people, for example, we become 'infected' by their nervousness. Emotional contagion occurs most obviously within crowds of people: at political rallies, at football games, and in concerts.

Empathy, Tania says, is different from emotional contagion in that it requires a distinction between another and ourselves: 'I know you are in pain and I know it is not mine,' she explains. And yet with empathy or empathic concern, we feel the other person's pain as if it were our own. Compassion, on the other hand, is a feeling of concern for someone else's wellbeing but without experiencing the same feelings as that person: 'I don't share your anger but I want you to feel better,' Tania says. Compassion is feeling *for* someone; empathy is feeling *as* someone.

Empathising with someone else's emotional experience activates the same neural networks that would be active if we were having this emotional experience. This is called 'affective resonance', which sounds to me like Daniel Siegel's concept of 'resonance circuits'.

Being with a depressed person who you care for leads to you feeling down too, Tania says. 'If you stay in empathy, you can get empathic distress.' Observing someone who is stressed changes our physiology: it raises cortisol levels, and changes heart rate and pupil size. Our mirror neurons fire to determine what the other person's actions mean. This gives us a second-hand experience of what it is like to be them.

Tania tells us that she first began her investigations on the neuroscience of empathy with Matthieu Ricard, the author of the book *Happiness: a guide to developing life's most important skill*, which I have already read. Ricard is a Tibetan Buddhist monk who, for most of his life, has actively practised compassion. Formerly he was a scientist, and he is interested in investigating the neurological correlates of contemplative practices.

Tania had him come to her lab. She put him into a functional MRI (fMRI) machine and asked him to 'resonate with the suffering and pain of others as if the pain is your own' — an instruction to be empathic. He did this for an hour. When she asked him what it was like to experience others' suffering, he said that he felt distressed. After this, he was keen to do his regular compassion practice to ease the distress. Tania requested that he do this while in the fMRI machine, which he did.

What Tania had seen on the fMRI scans during Ricard's empathy practice was the activation of the empathy-for-pain network: primarily, the anterior insula and the anterior medial cingulate cortex. But what she saw on the fMRI scans during Ricard's compassion practice was remarkable. A completely distinct network was activated: primarily, the medial orbitofrontal cortex, the pregenual anterior cingulate cortex, and the ventral striatum. From a neural point of view, she concluded, empathy and compassion are distinct emotional experiences.

What Tania says is an epiphany for me. It's suddenly clear to me that during my years of clinical work, each time I was with a client I was dipping into their pain — my body's physiology changing with theirs. I was experiencing their emotional suffering. When I was sitting face-to-face with the young woman's murderer, a man who feared for his own life, I was experiencing his fear too. During my Children's Court work, I was feeling the distress of the children I was with, and it was made worse by me imagining my own children in the same circumstances. I was undergoing, vicariously, the abuse and neglect of my children, in my body and in my brain.

During my psychology training, we were encouraged to attune to our clients by using communication micro skills: posture, positioning, the mirroring of body movements, and reading our own bodies to clue us into our clients' experience. We were trained in verbal skills called empathic and active listening: ways of checking in with our client to see if we were picking up correctly on their inner experience. These techniques helped us to fathom what it was like to be in the other person's world, to stand inside it with them. This means that psychologists, and other psychotherapists who are trained in a similar way, are being set up to experience empathic distress — the seed of vicarious trauma.

But Tania has good news. Because compassion is a different neural event from empathy, it allows for the option to retrain the brain to experience compassion rather than empathic distress.

Together with her PhD student, Olga Klimecki, Tania carried out a study with people who had no history of meditation or compassion practice. The experimental group was trained in empathy, and a second group was trained in a memory exercise for words (the control group). Before and after training, they were assessed on their responses to viewing, without sound, two types of video clips of real events. The low-emotion videos showed men, women, and children in everyday situations. The high-emotion videos were taken from news events and documentaries showing human suffering following injuries and natural disasters.

The empathy-trained subjects reported an increase in empathy and negative feelings after viewing both mildly and highly distressing videos, compared with the memory group. The empathy-trained group's fMRI scans showed activation of the empathy-for-pain network: the anterior insula and the anterior medial cingulate cortex.

Subsequently, the empathy-trained group was given instruction in a compassion practice: loving-kindness meditation, to engender feelings of warmth and care. This, as I knew by now, involved the visualisation of a close and loved person, a neutral person, and a difficult person, generating positive feelings towards them, and then projecting these feelings towards strangers and human beings in general.

Olga and Tania found that compassion training reversed the subjects' negative feelings after watching the video clips, reducing it back to baseline, and increased their positive feelings. Tania says these brain areas activated by compassion training — the medial orbitofrontal cortex, the pregenual anterior cingulate cortex, and the ventral striatum — are associated with the care, love, and reward systems and involve the neurotransmitters oxytocin, dopamine, and opioids. Oxytocin facilitates feelings of trust and love, dopamine gives a sense of reward, and opioids provide pain relief.

Tania thinks that compassion training may be a form of psychological inoculation for people faced with human distress or adverse circumstances. The compassionate orientation does not eliminate the experience of negative feelings. Rather, it seems to provide a re-interpretation of human suffering; it leads to the compassionate person experiencing increased warmth and concern for others.

She emphasises that this research is in an early phase. Further investigations will look at the effect of different methods of compassion training on different population groups.

I ask Tania about psychological trauma and what role mindfulness and meditation can play in helping sufferers to recover from trauma. I've been using these practices to help me recover from post-traumatic stress disorder, I tell her.

She reminds us that in psychological trauma, an automatic response is triggered by a perceived threat, and the inhibitory function of the prefrontal cortex doesn't work effectively during a state of high fear. Traditional therapeutic approaches, such as cognitive behaviour therapy and practising detachment or suppression of emotion, have a partial effect on dampening the trauma response, she says. Meditation, on the other hand, is 'something new'. She thinks it doesn't work on active inhibition, but on an earlier phase of not getting triggered to start with: when you 'stay with the emotion' but recognise that 'it's not me'. 'You don't need to put up a wall against it,' she says.

'Yes, it's not denial,' I say, 'it's staying with the emotion and knowing this. I still get a bodily reaction. I'm watching it, but I have a choice.'

'How long has it taken you to get to this stage, using mindfulness and meditation?' she asks.

'About two years,' I reply, 'but I don't have complete control.'

'That's short,' she says. 'Getting to this stage requires a lot of training, and it's not easy.'

She points out that the insula is the interoceptive cortex and acts as a relay station, bringing bodily sensations to the cortex, which interprets emotions based upon these sensations. Its activity is enhanced through mindfulness and meditation practices, and this may be one of the pathways that help to change the psychological response to traumatic memories. 'Training the heart' through loving-kindness practice is training the attachment system — the caregiving system that we respond to as babies and children. Mindfulness stabilises the mind so that we can more readily access our mental and sensory experiences, giving us the capacity to focus the mind in a way that alters brain function.

Tania has confirmed, so neatly, what I've been working out for myself: mindfulness and meditation have changed my brain for the better.

AFTER THE DAY is over, my mind tumbles with insights. I think of Choeying. She told me that she moved through empathy to compassion and that such compassion is necessary to be with others in distress. Does her version of compassion correlate with the activation of the compassion network that Tania has described?

Tania's preliminary findings suggest that those who employ a compassionate view of human distress are not dragged down by it. What I observed in Choeying was not detachment from those she was helping, but a desire to help, followed up by action. Choeying said that she felt uplifted in the face of others' suffering: not trying to fix their pain, but helping them to see a way through it, and, in the process, gaining insight.

Why was the Dalai Lama's teaching in Sydney such a positive experience for me? Was I responding, through the 'resonance circuits' that Daniel Siegel describes, to the presence of those around me — soaking up their calmness and their caring outlook? Was there something about the intense compassion I witnessed in the Dalai Lama that my body and my brain responded to?

As I look back over the five years since 2006, what stands out for me is the compassion of those who have helped me. My ocean-swimming friends, who were there each week to talk; Lily, who, through singing, showed me a way of transforming my heaviness; Wayne, my bedrock and counsel, helping me skilfully as a psychotherapist; Doctor Franklin, with his commonsense advice and advocacy for me with my insurer and others; Doctor Small, with his reassurance that I was doing the right things for my recovery, and encouraging me to write; my mates Ian, Doug, and David; James, with his encouragement of my neuroscience enquiries; Nick and my muso friends; my GP; and my former professional colleagues. I'm surprised to realise that many of these have been men — steadfast, caring, wise males who have been instrumental in my recovery. My faith in men has been restored.

And there is my family, who endured the intensity of my personal hell and the awful confluence of life events that beset us.

And me. When I walked into Wayne's office for the first time in October 2006, I was carrying failure, shame, and self-criticism, and an admission that I could not do this thing on my own. Now I have a feeling of renewal. Is this self-compassion?

I wonder: has compassion changed my brain? Tania's research suggests that it has.

My auditory processing has improved dramatically since the stroke — my comprehension of speech and my working memory in particular. I can think again, hold conversations, and make sense of what I hear and read. Mindfulness is no longer a mystical concept; it is a real, constant way of apprehending my inner world.

My emotional life has been rebalanced; is this due to the

assiduous activity of my prefrontal cortex? Writing and therapy have definitely helped. Exercise, nutrition, music, and forming nurturing relationships have amplified the neuroplastic changes.

I'm still not back to where I was, but I'm steadily improving. And in other ways, I'm far ahead of where I was. I've rescued my brain. Have I also found my psychological insurance policy?

IN THE YEAR's final session with Wayne, I ask, 'What are the long-term effects of post-traumatic stress disorder? Will I have a permanent disability?'

'You won't be the person you would have been if you had never been exposed to those experiences,' he says. 'Ask: *Am I still controlled by the trauma memories when they come unbidden?*'

He thinks I'm now in control, in many ways. When certain situations arise, like they did at the writers' festival, the original injury is like a scar. The old, disturbing memories are reawakened, and highly emotional when I access them. What's important is how well I manage them when this happens — how much it affects me in a day-to-day way.

'You will be particularly sensitive to children's stories,' he says. 'It's a healthy coping response to practise avoidance and denial at times. But if you live like some Vietnam veterans I've worked with, who exist by themselves in isolation, that is pathogenic; their experience is so limited they can never change the way they think about things.

'In your case, you are mixing with people and taking up new challenges, and your perspective on the way you think and react is changing. You've done your frontline duty and experienced the burnout that almost inevitably comes with this kind of work. You will not be the same person you would have been if you had never experienced these things, but you are wiser for it.'

I'm starting to like the new person.

EPILOGUE

RECENTLY, I WENT back to Lismore Hospital. I found the ward where I had stayed and walked the path I must have taken to the canteen. The hospital and rooms were smaller and duller than I remembered. The staff in the canteen that I had thought were 'at a party' looked like a regular bunch of workers on their break, this time. I didn't stop for a coffee.

A while ago, I gained my hospital records from my admission. These confirm that a CT scan was done. The CT report says: *No evidence of intracranial haemorrhage, infarction, or a space-occupying lesion.* I remember what Doctor Small had said about lesions sometimes not showing up on CT scans. I was seen by the serious doctor, a physician, in the afternoon of the day of my admission. He wrote in the notes: *Most likely TGA [transient global amnesia], with differential diagnoses of CVA [cerebrovascular accident] and encephalitis less likely.* I find it puzzling that throughout the notes amnesia is remarked upon, as well as confusion, but in some instances the notes say I was oriented to time and place. The notes reveal that the medical personnel were aware I had vomited on the way to hospital (reported by Anna) and that I had a headache. My C-reactive protein count, ascertained from the blood tests, was slightly elevated. These are all signs, I've since learnt, that are consistent with a stroke.

There is the serious doctor's written request for a psychiatric review, but no evidence of a psychiatrist or Doctor Banister coming in to see me before I was discharged.

Almost three years post-stroke, in March 2012, I undertook a full clinical review, with new brain MRI, an ultrasound, and blood tests, all ordered by the neuropsychiatrist I first came across at Seaview. Doctor Franklin had referred me to him for advice on the cause of my stroke and whether there was medication I could take to help with my neurological condition. The neuropsychiatrist found no medical reason for me to have had the stroke. Although he agreed that there is a link between depression and stroke, he said that the underlying mechanism for this has not been delineated. My arteries were clear, and there was no evidence of further bleeding in the brain.

He also confirmed that mental fatigue is common following brain injury, but could not explain why; he suggested a sleep study to ascertain if sleep apnoea was a cause of my mental fatigue, but this revealed no sleep disorder. Yet I was relieved that he confirmed I was doing all I could for my rehabilitation, with no further medication appropriate other than the 100 milligrams of aspirin directed by Doctor Small and the fish-oil tablets. He thought that Doctor Small had done all the right things.

The neuropsychiatrist referred me to a neurologist. He looked at all three sets of MRI scans done over the three years and gave me a more detailed description of the likely areas of damage to my brain. His observations confirmed that areas in my left temporal lobe were probably more affected than suggested in the radiologist's initial MRI report (which is what had led to the stroke diagnosis), and the 'tail' of my left hippocampus had been damaged. This makes sense to me because of the constellation of auditory, memory, geographical, and learning deficits I experienced — these were more extensive than would be expected solely from an occipital infarction. The neurologist said that there would have been more brain swelling evident on the day of the stroke, and soon afterwards, than appears on the first MRI, which was done three weeks post-stroke.

Just over four years post-stroke, in September 2013, my ophthalmologist, Doctor Mercer, conducted a visual-field test. During this, I was able to see pinpoints of light in the previously dark upper-right quadrant, although they were fainter than the points of light in the other parts of my visual field. Doctor Mercer declared that my visual capacity had returned to normal, nevertheless.

Recently, I started cognitive training using the latest Posit Science exercises for auditory processing, now called BrainHQ. I was pleased to see that I moved through the basic levels of these new exercises very quickly, and I am now on to the harder levels.

All my treatment practitioners have asserted that I should not return to clinical psychological work; it remains too risky. In a medical sense, I have not completely recovered, neurologically or psychologically. However, I feel recovered, in a very important way: I very much like the person I am now. I have strengths that I didn't have prior to the stroke, and I accept that some of the old strengths, such as the analytical mind and the sharp memory, are gone.

My intention is to continue with writing, public speaking, and advocacy for mental-health and disability issues. Oh, and with drinking coffee, playing music, being a father, swimming — and investigating my brain.

Wayne has since retired from clinical practice and is engaged in other pursuits. Choeying has withdrawn from formal teaching and no longer lives in Hervey Bay.

FURTHER READING

THERE HAVE BEEN several books, people, and organisations that have helped me in my recovery from trauma and stroke between 2006 and 2011. The following is a list of the principal resources I read or accessed during this time. I have included them, along with some other references, for those who would like to read more about these topics. They may be helpful to those going through similar experiences to mine.

Stroke and brain injury

Carter, Rita, *The Human Brain Book: an illustrated guide to its structure, function, and disorders*, Dorling Kindersley, London, 2009.

The MRI scans of a real brain, and the graphics showing brain function, in this wonderful illustrated guide enabled me to visualise my brain and get a better sense of how it worked (or didn't). It is a terrific starting point for understanding brain anatomy and function, and would be particularly useful for those who have trouble with reading, as the images are plentiful.

National Stroke Foundation, www.strokefoundation.com.au, and Brain Injury Australia, www.bia.net.au

These websites, which I referred to frequently after my stroke, provide easy-to-understand information about the effects of brain injury. They have links to other useful services and organisations.

Osborn, Claudia L., *Over My Head: a doctor's own story of head injury from the inside looking out*, Andrews McMeel Publishing, Kansas City, 2000.

This memoir describes the author's experience of life after a head injury, and how she approached her cognitive recovery. It confirmed for me that the cognitive difficulties I was experiencing were real, and gave me more encouragement to address them.

Taylor, Jill Bolte, *My Stroke of Insight: a brain scientist's personal journey*, First Plume Printing, New York, 2009, and 'My Stroke of Insight', *TED*, 28 February 2008, www.ted.com/talks/jill_bolte_taylor_s_powerful_stroke_of_insight

Taylor's book was the first account I read of another's stroke. I did not understand the full import of her message until I watched her TED talk: it was her description of a 'nirvana-like' experience following her stroke that confirmed for me that my feeling of transcendence during my stroke had a neurological basis. I recommend watching the talk if you have had a stroke or know someone who has. It's thought-provoking stuff.

Trauma

Saakvitne, Karen and Pearlman, Laurie, *Transforming the Pain: a workbook on vicarious traumatization*, W.W. Norton & Company, New York, 1996, and Skovholt, Thomas and Trotter-Mathison, Michelle, *The Resilient Practitioner: burnout prevention and self-care strategies for counselors, therapists, teachers, and health professionals*, Allyn & Bacon, Boston, 2001.

These books helped me to understand how my psychology work had led me to suffer emotional damage. Reading them gave me the impetus to seek out the help of a psychologist (Wayne).

The Australian Centre for Posttraumatic Mental Health, www.acpmh.unimelb.edu.au

This research centre, based at the University of Melbourne, offers information for those suffering from psychological trauma and for treating professionals.

Beyond Blue, www.beyondblue.org.au, and the Mental Health Foundation, www.mentalhealth.org.uk

Beyond Blue provides information and tools relevant to the diagnosis and treatment of depression, anxiety, and trauma. The Mental Health Foundation, a UK-based charity, covers all aspects of mental health. It also offers an online course in mindfulness.

Mindfulness and compassion

Chodron, Pema, *When Things Fall Apart: heart advice for difficult times*, HarperCollins, London, 1997.

In this beautifully practical book, Buddhist teacher and nun Pema Chodron argues that life is inherently insecure and it is best to make peace with this reality. When it felt like everything in my life was falling apart — psychologically and financially — this book gave me solace, and the courage to confront fear and pain.

Gilbert, Paul and Choden, *Mindful Compassion: using the power of mindfulness and compassion to transform our lives*, Constable Robinson, London, 2013.

The therapeutic application of compassion is an area that James Bennett-Levy and I are particularly interested in. Paul Gilbert, a professor of psychology, is the founder of compassion-focused therapy, which draws upon evolutionary psychology, neuroscience, western psychotherapy, mindfulness, and compassion practices.

Ricard, Matthieu, *Happiness: a guide to developing life's most important skill*, Atlantic Books, London, 2007.

Matthieu Ricard's assertion that happiness does not ultimately come from seeking the fulfilment of personal desire struck a chord with me, as it might with others. Ricard was the Buddhist monk who assisted Tania Singer in her discovery of the difference between the empathic and compassionate response in the brain.

Segal, Zindel; Williams, Mark; and Teasdale, John, *Mindfulness-Based Cognitive Therapy for Depression: a new approach to preventing relapse*, Guildford Press, New York, 2002.

I turned to this book when I was seeking a 'psychological insurance policy'; it reminded me that mindfulness can be useful for emotional resilience. It features practical exercises and a theoretical explanation of mindfulness and cognitive behaviour therapy. While written for the mental-health professional, lay readers can also take much from it.

Singer, Tania and Bolz, Matthias (eds), *Compassion: bridging practice and science*, Max Planck Institute for Human Cognitive and Brain Sciences, Leipzig, 2013, www.compassion-training.org

This ebook collates multidisciplinary and multimedia presentations from a conference titled 'How to Train Compassion', which was held in Berlin, Germany, in July 2011. Available free online, it is useful for those interested in the science behind compassion, and how to train others in compassion and meditation.

The Center for Mindful Self-Compassion, www.centerformsc.org

This website provides guided exercises, links, and trainings for those interested in the mindful self-compassion approach developed by Kristin Neff and Chris Germer.

Tobler, Albert and Herrmann, Susann, *The Rough Guide to Mindfulness: the essential companion to personal growth*, Rough Guides, London, 2013.

This publication gives a comprehensive explanation of how mindfulness works and how it can be incorporated into many aspects of life. It would be useful both for those finding their way into this topic and for those who have been working with mindfulness for a while.

Neuroscience

Arden, John and Linford, Lloyd, *Brain-Based Therapy with Adults: evidence-based treatment for everyday practice*, John Wiley & Sons, Hoboken, 2009, and Cozolino, Louis, *The Neuroscience of Psychotherapy: healing the social brain*, 2nd edition, W. W. Norton & Company, New York, 2010.

These books, together with Siegel's works, were the main titles I read on the application of brain science to psychotherapy. They are perhaps more academic in tone than most of the books on this list, but they have useful insights to offer the keen reader.

Davidson, Richard (with Begley, Sharon), *The Emotional Life of Your Brain: how its unique patterns affect the way you think, feel and live — and how you can change them*, Hodder, London, 2012.

Richard Davidson has spent his career investigating the neuroscience of emotion, carrying out groundbreaking research on meditation. His work led him to conceive of six dimensions of emotional styles, and in this book he suggests ways of moderating these styles through the application of neuroscientific principles.

Doidge, Norman, *The Brain That Changes Itself: stories of personal triumph from the frontiers of brain science*, Scribe Publications, Melbourne, 2007.

This eye-opening book introduced the notion of neuroplasticity to many readers, including me. It is written for the general reader, so I was capable of understanding it after my stroke. It gave me significant insight into how I might be able to aid my cognitive and emotional recovery.

Hanson, Rick (with Mendius, Richard), *Buddha's Brain: the practical neuroscience of happiness, love, and wisdom*, New Harbinger Publications, Oakland, 2009.

This book stimulated my thinking about the neuroscience of contemplative states of mind and the idea that our sense of self is a construct. Referencing reliable research, it confirmed for me that meditation and mindfulness change the brain for the better.

Klimecki, Olga et al., 'Differential Pattern of Functional Brain Plasticity After Compassion and Empathy Training', *Social Cognitive and Affective Neuroscience*, 10 April 2013, www.scan.oxfordjournals.org/content/early/2013/05/09/scan.nst060.full, and Klimecki, Olga et al., 'Functional Neural Plasticity and Associated Changes in Positive Affect After Compassion Training', *Cerebral Cortex*, vol. 23, no. 7, pp. 1552–61.

These references describe the research that I first heard Tania Singer speak about in James Bennett-Levy's living room. Singer suggests that a compassionate outlook might be a better way for health professionals to stay resilient when facing human distress.

Siegel, Daniel J., *The Mindful Brain: reflection and attunement in the cultivation of wellbeing*, W. W. Norton & Company, New York, 2007, and *Mindsight: change your brain and your life*, Scribe Publications, Melbourne, 2009.

Daniel Siegel's books reinforced for me that brain science could be applied to therapy for mental-health disorders. His assertion that mindfulness meditation strengthens the prefrontal cortex's ability to regulate emotions motivated me to keep practising these techniques.

The Mindsight Institute, www.mindsightinstitute.com

This is the website set up by Daniel Siegel. It offers videos of Siegel and access to his online courses.

ACKNOWLEDGEMENTS

I WISH TO express my gratitude to the Northern Rivers Writers' Centre; it presented me with a world of writing almost at my door through workshops, consultants, and an annual writers' festival. The staff were always ready to help. During the period I was working on this memoir, those who assisted me included Susie, Siboney, Penny, and Sarah.

Through the writers' centre I had consultations with Peter Bishop, Irina Dunn, Marele Day, and Laurel Cohn, each of whom asked me pointed questions, and gave reassurance and advice until I came to believe that I had a potential 'book in me'. However, it was through Alan Close, my early mentor, that I received the greatest encouragement. I treasure the meetings we had, as he listened intently to my outpourings and one repeated question: 'Do you think *that* should be in the book?'

'That's good. Put it in,' he always said.

Alan encouraged me to write whatever came to mind, and I did, even though many of the scenes I wrote did not make it into the final draft; but I needed to get them out before the narrative arc of *How I Rescued My Brain* revealed itself.

However, after Alan's mentorship, I reached a low point, wondering if I could ever write sufficiently well to be published. Fortunately, I contacted Jesse Blackadder, an accomplished local author, who over lunch in January 2012 gave me three pieces of advice. I followed this advice, and later that year I was awarded a fellowship at Varuna, the National Writers' House, in Katoomba, where I was able to complete most of the first draft. That same year I pitched my book at the Byron Bay Writers Festival, garnering sufficient interest to secure an agent, Gaby Naher, and through her assiduous efforts a publication contract shortly afterwards.

I wish to thank the teachers and students I encountered over the duration of three online memoir-writing courses with Gotham Writers' Workshop in 2012 and 2013. When people from other countries gave advice or praise, it gave me the sense that what I was writing about had resonance. Through their guidance I learnt many of the skills of writing narrative nonfiction, as well as what scenes to expand or shrink, and how to address specific problems in the pieces I submitted for comment.

I also wish to thank Diana, Eddy, Jill, Sally, and Prem from my memoir-writing group, who, most memorably, exhorted me to put more emotion on the page (even when I thought I already was!). I needed to shed my academic-writing mindset. I hope they are happy with the result.

I was also awarded a LitLink Residential Fellowship at Varuna in late 2013. This enabled me to address structural issues raised by my editor, Julia Carlomagno, and to capture and put onto the page subtler internal experiences that I had passed over in the first draft. The environment at Varuna allowed me to reflect, to concentrate, and to be liberated from everyday concerns so that I could stay immersed in the story. I want to thank the other writers I met during the Varuna residencies. I enjoyed our lively and humorous discussions. And I wish to thank the staff: Jansis, Sheila, Vera, and Joan, who were finely attuned to the sensibilities of the writing life.

I am thankful for comment from several people on specific aspects of the manuscript. These individuals include Alan Close, Gaby Naher, Sharon Dean, Claire Dunn, James Bennett-Levy, Jan Maehl, and also Laurel Cohn, who did a close edit of my first three chapters before I submitted them to the publisher. To the many characters in this story who helped me to recreate specific scenes, your contributions have made it a richer record.

I met Claire Dunn during my first Varuna residency, when she was completing her memoir *My Year Without Matches*. We were at similar stages in our works and facing similar challenges, and it has been wonderful to have someone to 'hold hands with' along the path to publication.

When my friend Dix saw me working in my cramped and cluttered writing space at home, during the critical editing phase, he swiftly brought about a reorganisation, transforming it into an uncluttered and airy space that I looked forward to entering in the mornings.

I want to thank my publisher, Scribe. I made the right choice to go with this publishing house, and I've felt supported throughout the editing, design, and publicity stages. My working relationship with my editor, Julia, has provided me with an unexpected writing education. She has exhibited great respect towards me, my writing method, and my ideas. While I have had my hand on the tiller of this particular vessel, she has trimmed the sails, making the final manuscript considerably more enjoyable for the reader, I'm sure. I also wish to thank the art director, Miriam Rosenbloom, and the cover designer, Allison Colpoys, for their work. The cover features *kintsukuroi* pottery. The Japanese concept of *kintsukuroi* refers to the repair of pottery with gold or silver lacquer, with the understanding that the piece is more beautiful for having been broken, which is a lovely metaphor.

I want, too, to thank my immediate family. Firstly, my daughters, who thought that I went on a little too much about my brain: I hope this book provides you with insights you could not have gained when we were going through our ordeals. Secondly, my ex-wife, who, while understandably finding some of the material in the book exposing, gave me feedback that made it a better account. Thirdly, my siblings, for their love and support. And lastly, the many friends who have been encouraging of 'the book', as it became known among us, listening to my interminable updates on progress.

And I wish to thank you, the reader. Sometimes, when I was in a pit of despair and searching for a reason to make the effort to keep going, I would tell myself, *I have to get out of this and write about it, and perhaps in doing so give hope to others going through a similar experience.* Though I don't know you, you were there in my thoughts then, encouraging me. Thank you.